Adaptive Motion Compensation in Radiotherapy

IMAGING IN MEDICAL DIAGNOSIS AND THERAPY

William R. Hendee, Series Editor

Quality and Safety in Radiotherapy
Todd Pawlicki, Peter B. Dunscombe, Arno J. Mundt, and
Pierre Scalliet, Editors
ISBN: 978-1-4398-0436-0

Adaptive Radiation Therapy
X. Allen Li, Editor
ISBN: 978-1-4398-1634-9

Quantitative MRI in Cancer
Thomas E. Yankeelov, David R. Pickens, and
Ronald R. Price, Editors
ISBN: 978-1-4398-2057-5

Informatics in Medical Imaging
George C. Kagadis and Steve G. Langer, Editors
ISBN: 978-1-4398-3124-3

Adaptive Motion Compensation in Radiotherapy
Martin J. Murphy, Editor
ISBN: 978-1-4398-2193-0

Forthcoming titles in the series

Image-Guided Radiation Therapy
Daniel J. Bourland, Editor
ISBN: 978-1-4398-0273-1

Informatics in Radiation Oncology
Bruce H. Curran and George Starkschall, Editors
ISBN: 978-1-4398-2582-2

Image Processing in Radiation Therapy
Kristy Kay Brock, Editor
ISBN: 978-1-4398-3017-8

Proton and Carbon Ion Therapy
Charlie C.-M. Ma and Tony Lomax, Editors
ISBN: 978-1-4398-1607-3

Monte Carlo Techniques in Radiation Therapy
Joao Seco and Frank Verhaegen, Editors
ISBN: 978-1-4398-1875-6

Stereotactic Radiosurgery and Radiotherapy
Stanley H. Benedict, Brian D. Kavanagh, and
David J. Schlesinger, Editors
ISBN: 978-1-4398-4197-6

Cone Beam Computed Tomography
Chris C. Shaw, Editor
ISBN: 978-1-4398-4626-1

Handbook of Brachytherapy
Jack Venselaar, Dimos Baltas, Peter J. Hoskin, and
Ali Soleimani-Meigooni, Editors
ISBN: 978-1-4398-4498-4

Targeted Molecular Imaging
Michael J. Welch and William C. Eckelman, Editors
ISBN: 978-1-4398-4195-0

IMAGING IN MEDICAL DIAGNOSIS AND THERAPY

William R. Hendee, Series Editor

Adaptive Motion Compensation in Radiotherapy

Edited by
Martin J. Murphy

CRC Press
Taylor & Francis Group
Boca Raton London New York

CRC Press is an imprint of the
Taylor & Francis Group, an **informa** business

CRC Press
Taylor & Francis Group
6000 Broken Sound Parkway NW, Suite 300
Boca Raton, FL 33487-2742

First issued in paperback 2018

© 2012 by Taylor & Francis Group, LLC
CRC Press is an imprint of Taylor & Francis Group, an Informa business

No claim to original U.S. Government works

ISBN-13: 978-1-4398-2193-0 (hbk)
ISBN-13: 978-1-138-37429-4 (pbk)

Contents

Series Preface vii

Preface ix

About the Editor xi

List of Contributors xiii

Introduction xv
  Martin J. Murphy

1 Real-Time Tumor Localization 1
  Ruijiang Li, Laura I. Cerviño, and Steve B. Jiang

2 Theoretical Aspects of Target Detection and Tracking 13
  Gregory C. Sharp, Rui Li, and Nagarajan Kandasamy

3 Respiratory Gating 21
  Geoffrey D. Hugo and Martin J. Murphy

4 The CyberKnife® Image-Guided Radiosurgery System 29
  Martin J. Murphy

5 Fundamentals of Tracking with a Linac Multileaf Collimator 33
  Dualta McQuaid and Steve Webb

6 Couch-Based Target Alignment 39
  Kathleen T. Malinowski and Warren D. D'Souza

7 Robotic LINAC Tracking Based on Correlation and Prediction 47
  Floris Ernst and Achim Schweikard

8 Treatment Planning for Motion Adaptation in Radiation Therapy ... 65
  Alexander Schlaefer

9 Treatment Planning for Motion Management via DMLC Tracking ... 77
  Lech Papiez and Dharanipathy Rangaraj

10 Real-Time Motion Adaptation in Tomotherapy® Using a Binary MLC ... 93
  Weiguo Lu, Mingli Chen, Carl J. Mauer, and Gustavo H. Olivera

11 Combination of a LINAC with 1.5 T MRI for Real-Time Image Guided Radiotherapy ... 107
  Jan J.W. Lagendijk, Bas W. Raaymakers, and Marco van Vulpen

12 The ViewRay™ System 115
  Daniel A. Low, Richard Stark, and James F. Dempsey

13 Fault Detection in Image-Based Tracking 129
  Gregory C. Sharp, Rui Li, and Nagarajan Kandasamy

Index 135

Series Preface

Advances in the science and technology of medical imaging and radiation therapy are more profound and rapid than ever before, since their inception over a century ago. Further, the disciplines are increasingly cross-linked as imaging methods become more widely used to plan, guide, monitor, and assess treatments in radiation therapy. Today, the technologies of medical imaging and radiation therapy are so complex and so computer-driven that it is difficult for the persons (physicians and technologists) responsible for their clinical use to know exactly what is happening at the point of care, when a patient is being examined or treated. The persons best equipped to understand the technologies and their applications are medical physicists, and these individuals are assuming greater responsibilities in the clinical arena to ensure that what is intended for the patient is actually delivered in a safe and effective manner.

The growing responsibilities of medical physicists in the clinical arenas of medical imaging and radiation therapy are not without their challenges, however. Most medical physicists are knowledgeable in either radiation therapy or medical imaging and expert in one or a small number of areas within their discipline. They sustain their expertise in these areas by reading scientific articles and attending scientific lectures at meetings. In contrast, their responsibilities increasingly extend beyond their specific areas of expertise. To meet these responsibilities, medical physicists periodically must refresh their knowledge of advances in medical imaging or radiation therapy, and they must be prepared to function at the intersection of these two fields. How to accomplish these objectives is a challenge.

At the 2007 annual meeting of the American Association of Physicists in Medicine in Minneapolis, this challenge was the topic of conversation during a lunch hosted by Taylor & Francis Group among a group of senior medical physicists (Arthur L. Boyer, Joseph O. Deasy, C.-M. Charlie Ma, Todd A. Pawlicki, Ervin B. Podgorsak, Elke Reitzel, Anthony B. Wolbarst, and Ellen D. Yorke). The conclusion of this discussion was that a book series should be launched under the Taylor & Francis banner, with each volume in the series addressing a rapidly advancing area of medical imaging or radiation therapy of importance to medical physicists. The aim for each volume would be to provide medical physicists with the information needed to understand technologies driving a rapid advancement and their applications to safe and effective delivery of patient care.

Each volume in the series is edited by one or more individuals with recognized expertise in the technological area encompassed by the book. The editors are responsible for selecting the authors of individual chapters and ensuring that the chapters are comprehensive and intelligible to someone without such expertise. The enthusiasm of volume editors and chapter authors has been gratifying and reinforces the conclusion of the Minneapolis luncheon that this series of books addresses a major need of medical physicists.

Imaging in Medical Diagnosis and Therapy would not have been possible without the encouragement and support of the series manager, Ms. Luna Han of Taylor & Francis Group. The editors and authors, and most of all I, are indebted to her steady guidance of the entire project.

William R. Hendee
Series Editor
Rochester, MN

Preface

External-beam radiotherapy has long been vexed by the simple fact that patients can (and do) move during the delivery of radiation. The most elegant and forward-looking solution to this reality is to actively adapt the radiation delivery process to the patient's natural movements. Recent advances in imaging and beam delivery technologies have now made this solution a practical reality. The purpose of this book is to present to researchers and clinical practitioners in radiation therapy an overview of the current and prospective state of the art in motion-adaptive radiation therapy. It presents technical reviews of each of the contributing elements of a motion-adaptive system (including target detection and tracking, beam adaptation, and patient realignment), discusses treatment planning issues that arise when the patient and internal target are mobile, describes several integrated motion-adaptive systems that are in clinical use or at advanced stages of development, and concludes with a review of the system control functions that must be an essential part of any therapy device that operates in a near-autonomous manner with limited human interaction. From these chapters, the reader will hopefully gain not only an understanding of the technical aspects and capabilities of motion adaptation but also practical clinical insights into planning and carrying out various types of motion-adaptive radiotherapy treatment.

MATLAB® is a registered trademark of The MathWorks, Inc. For product information, please contact:

The MathWorks, Inc.
3 Apple Hill Drive
Natick, MA 01760-2098 USA
Tel: 508 647 7000
Fax: 508-647-7001
E-mail: info@mathworks.com
Web: www.mathworks.com

About the Editor

Dr. Martin J. Murphy received his PhD in physics from the University of Chicago in 1980. Following postdoctoral fellowships in nuclear physics at the University of California/Berkeley and the University of Washington and a stint as a research scientist in gamma-ray astronomy at the Lockheed Palo Alto Research Laboratories, he entered the field of radiation therapy research and development in 1992 as Director of System Development of the CyberKnife® at Accuray Incorporated. In 1995, he joined the Department of Radiation Oncology at Stanford University as a senior research scientist to continue development of the CyberKnife's image guidance and target tracking capabilities. In 2003, Dr Murphy joined the Department of Radiation Oncology at Virginia Commonwealth University, where he is presently engaged in several research programs involving medical image registration, CT reconstruction, and real-time motion-adaptive control systems.

List of Contributors

Laura I. Cerviño, PhD
Department of Radiation Oncology
University of California San Diego
Moores Cancer Center
La Jolla, California

Mingli Chen, PhD
TomoTherapy Inc.
Madison, Wisconsin

James F. Dempsey, PhD
ViewRay™, Inc.
Oakwood Village, Ohio

Warren D. D'Souza, PhD
Department of Radiation Oncology
University of Maryland School of
    Medicine,
Baltimore, Maryland
and
Fischell Department of
    Bioengineering
A. James Clark School of Engineering
University of Maryland
College Park, Maryland

Floris Ernst, PhD
Institute for Robotics and Cognitive
    Systems
University of Lübeck
Lübeck, Germany

Geoffrey D. Hugo, PhD
Department of Radiation Oncology
Virginia Commonwealth University
Richmond, Virginia

Steve B. Jiang, PhD
Department of Radiation Oncology
University of California San Diego
Moores Cancer Center
La Jolla, California

Nagarajan Kandasamy, PhD
Department of Electrical and Computer
    Engineering,
Drexel University,
Philadelphia, Pennsylvania

Jan J.W. Lagendijk, PhD
Department of Radiotherapy
University Medical Centre Utrecht
Utrecht, the Netherlands

Rui Li, PhD
Department of Radiation Oncology
Massachusetts General Hospital
Boston, Massachusetts

Ruijiang Li, PhD
Department of Radiation Oncology
University of California San Diego
Moores Cancer Center
La Jolla, California

Daniel A. Low, PhD
Department of Radiation Oncology
UCLA
Los Angeles, California

Weiguo Lu, PhD
TomoTherapy Inc.
Madison, Wisconsin

Kathleen T. Malinowski, MS
Department of Radiation Oncology
University of Maryland School of Medicine
Baltimore, Maryland
and
Fischell Department of
    Bioengineering
A. James Clark School of Engineering
University of Maryland
College Park, Maryland

Carl J. Mauer, MS
TomoTherapy Inc.
Madison, Wisconsin

Dualta McQuaid, PhD
Francis H Burr Proton Therapy Center
Massachusetts General Hospital
Boston, Massachusetts

Martin J. Murphy, PhD
Department of Radiation Oncology
Virginia Commonwealth University
Richmond, Virginia

Gustavo H. Olivera, PhD
TomoTherapy Inc.
Madison, Wisconsin

Lech Papiez, PhD
Department of Radiation Oncology
University of Texas Southwestern
    Medical Center
Dallas, Texas

Bas W. Raaymakers, PhD
Department of Radiotherapy
University Medical Centre Utrecht
Utrecht, the Netherlands

Dharanipathy Rangaraj, PhD
Department of Radiation Oncology
Washington University School of
    Medicine
Saint Louis, Missouri

Alexander Schlaefer, PhD
Medical Robotics Group
Institute for Robotics and Cognitive
    Systems
University of Lübeck
Lübeck, Germany

Achim Schweikard, PhD
Institute for Robotics and Cognitive
  Systems
University of Lübeck
Lübeck, Germany

Gregory C. Sharp, PhD
Department of Radiation Oncology
Massachusetts General Hospital
Boston, Massachusetts

Richard Stark, MS
ViewRay™, Inc.,
Oakwood Village, Ohio

Marco van Vulpen, MD, PhD
Department of Radiotherapy
University Medical Centre Utrecht
Utrecht, the Netherlands

Steve Webb, PhD, DSc
Joint Department of Physics
Institute of Cancer Research and Royal
  Marsden NHS Foundation Trust
London, UK

Introduction

Martin J. Murphy

Until recently, external-beam radiotherapy was obliged to utilize linear accelerator systems with fixed beam isocenters and stationary couches. Treatment delivery was developed as a static process in which the intended treatment plan would be delivered only if the tumor remained stationary at the beam isocenter. If the tumor moved relative to that isocenter, the delivered dose would change. Dealing with target movement in a static delivery paradigm requires *accommodating* the likely movement with various prospective planning strategies. Generally, this has consisted of planning an expanded dose distribution margin that encompasses the likely range of intrafraction target movement. A well-designed margin will deliver the intended dose to the target as it moves around but at the cost of increased radiation dose to neighboring tissues. Also, being a prospective compensation strategy, the margin approach must anticipate the likely movement before it happens. An inaccurate projection of the range of movement will result in an inaccurate dose delivery.

Recent technological advancements have enabled dynamic dose delivery in which the delivery system responds to movement in real time with a corrective response that maintains the equivalent of a fixed beam/target isocenter. In these new scenarios, static accommodation is replaced by dynamic adaptation. The subject of this book is the present state of the art in dynamic motion-adaptive external-beam radiotherapy.

Treatment sites can move under a variety of physiological influences, both stochastic and deterministic. Responding to movement with an adaptive response takes a certain amount of time. Time delays between the determination of the tumor position and the adaptive response must be handled by predicting where the tumor will be when the correction is effected. As a consequence, motion-adaptive systems adopt one of two strategies, depending on the nature of the movement. Deterministic movement, such as breathing, follows a trajectory that can be tracked such that the position of the tumor is known more or less continuously. A deterministic trajectory can be predicted with a reasonable degree of confidence and accuracy. In this situation, it is possible in principle to either move the beam synchronously with the tumor or gate the beam on and off as the tumor moves in and out of the line of sight. Random movements, which are inherently unpredictable, cannot be effectively tracked or gated; in this case, the adaptive system can (at best) make periodic samples of the tumor position at a rate that keeps the average position uncertainty below some prescribed level.

Adaptation to random movement is essentially a reactive process that does not provide much opportunity for strategically planning the treatment ahead of time. Deterministic (i.e.,

regular, systematic, and periodic) movements, on the other hand, allow the opportunity to proactively design and plan around the expected motion. Breathing has become the paradigmatic example of deterministic movement; thus adaptation to it has become the primary objective of motion-adaptive radiotherapy. Consequently, beam gating and tracking in response to respiration occupies a central place in all of the discussions in this book.

The process of adapting to deterministic target motion during radiation therapy can be divided into four tasks: (1) observing and assessing the anatomical motion before treatment for the purposes of planning; (2) devising a treatment plan that accommodates movement and takes advantage of it whenever possible; (3) identifying the precise position of the tumor at each moment in time during treatment delivery; and (4) performing a corrective beam alignment response synchronously with the motion in a way that is coordinated with the treatment plan. Additionally, a complete motion-adaptive radiotherapy system must have a real-time control loop that links the target acquisition system to the beam adaptation system, and must perform robust fault detection procedures to ensure that the system always behaves in a safe, appropriate, and effective manner. This book will address all of these issues in the context of a variety of adaptation strategies.

To be effective, an adaptive system must be able to respond in real time to motion. Real-time adaptation can be defined as a response time that is short compared to the characteristic time it takes the tumor to move a significant distance. For the sake of argument, a significant distance might be taken to be 2 mm. During normal quiet breathing, a lung tumor can move at velocities of 5–25 mm/s. This implies that a real-time respiration tracking system should be able to update the tumor's position at approximately 100-ms intervals, which requires a nominal position acquisition frame rate of about 10 Hz. However, given the quasiperiodic nature of breathing, a lower sampling frequency can give a reasonable approximation to a moving target's trajectory.

The real-time tracking frame rate is not the same as the total delay, or lag time, of the adaptive system. Processing image or other tracking data captured at 10 Hz, extracting the target position, and completing the system's response can take considerably longer than 100 ms. Conversely, measurements made at 2 Hz might be processed and corrected with a lag time of less than 100 ms. The link between the target tracking subsystem and the beam adaptation subsystem is a control and feedback loop that must, among other things, anticipate the future position of a moving tumor to compensate system lag time (latency). For respiratory motion, this requires predicting the patient's breathing behavior anywhere from 100 ms to 1000 ms in advance. Although breathing is superficially periodic, it exhibits a variety of nonstationary behaviors that have made breathing prediction a subject of ongoing research in motion-adaptive radiotherapy.

This book provides an overview of the following fundamental aspects of motion adaptation: (1) target detection and tracking; (2) mechanisms of adaptive system response; (3) treatment planning in the presence of anatomical movement; and (4) verification and quality assurance during the treatment process.

In the first chapter, Li, Cerviño, and Jiang address the practical problem of finding and following a moving treatment target. This requires either seeing the target directly via an imaging or tracking device or inferring its location by monitoring surrogate markers whose movement is highly and predictably correlated with the tumor position. Whenever possible, one would like to observe the tumor itself, but in the majority of situations, this isn't possible. However, if one can implant markers in the tumor that are either visible in x-ray images or that can be located electromagnetically, then one can achieve nearly direct tumor tracking. For radiographically-obscure tumors that cannot be marked by implanted fiducials, it becomes necessary to find a secondary observable that can predict for the tumor position. For example, a lung tumor moving during respiration might have a well-correlated relationship with movement of the chest surface, which can be monitored with TV cameras, infrared emitters, or other means. Finding the best surrogate for a particular tumor's motion remains an active area of research.

Following the presentation of localization technologies and techniques, Sharp presents in Chapter 2 a discussion of some theoretical aspects of target detection and tracking, paying particular attention to image-based tracking of fiducials and target anatomy. This establishes a theoretical basis for addressing the problems of reliability and confidence in target localization, which are addressed in a later chapter on fault tolerance and detection.

Once the tumor position has been acquired, there are three possible adaptive responses: (1) turn the beam on or off (beam gating) depending on the tumor's proximity to the treatment isocenter; (2) shift the beam itself to the present tumor position; or (3) move the patient such that the tumor stays at a position directly in line with the (fixed) beam position.

At the present time, beam gating is the most commonly employed form of motion adaptation, primarily because existing radiotherapy accelerators can be retrofitted for this function without great expense. Gating is reviewed by Hugo and Murphy in Chapter 3.

For external x-ray beam radiotherapy, adapting the beam and target alignment can be effected in three ways. The linear accelerator itself can be moved such that the beam continually points at the current tumor position. This method is clinically implemented in the CyberKnife®, which is currently the only clinical system capable of real-time respiratory tracking. Chapter 4 presents a brief technical description of the CyberKnife®,

with emphasis on the features that are specifically designed for tracking respiratory motion. For fixed-gantry accelerator systems equipped with a multileaf collimator (MLC), the collimated aperture can be programmed to move synchronously with tumor movement that is in the same plane as the aperture. McQuaid and Webb present the practical aspects of MLC adaptation in Chapter 5. In Chapter 6, D'Souza describes a highly innovative process for dynamically shifting the couch such that the tumor movement is offset and the tumor treatment isocenter remains fixed at the treatment beam isocenter. (For the case of charged particle therapy beams, it is possible to steer the beam electromagnetically in response to tumor motion, but this special situation is outside the scope of this book.)

The connection between the target detection system and the beam adaptation system is made via a real-time control loop. This loop must translate the observed tracking signal (which might be the actual tumor position or the movement of a respiratory surrogate) into the beam's new targeting coordinates, while making allowances for control-loop delays. Ernst and Schweikard present several current methodologies for performing this function in Chapter 7. Although their techniques have been specifically designed for the CyberKnife respiratory tracking system, they are fully generalizable to any other motion-adaptive radiation therapy system.

A moving target presents special treatment planning optimization problems and opportunities. Alexander Schlaefer presents some methods to approach these problems using a robotically-manipulated linear accelerator (e.g., the CyberKnife') in Chapter 8. In Chapter 9, Rangaraj and Papiez present the mathematical formalism for deriving MLC leaf sequences that include the response to a moving target. In particular, they show how the relative motion of the tumor with respect to normal tissue and critical structures presents an opportunity to exploit the shifting position of the target to reduce healthy tissue exposure.

One existing external-beam radiotherapy system and two more under development have highly integrated imaging and beam delivery technologies that provide unique methodologies for motion adaptation. Because of their distinctive configurations, each is described in a separate chapter. The Tomotherapy system, which integrates x-ray tomographic (i.e., CT) imaging and dose delivery, is described in Chapter 10 by Luo et al. This system uses radiographic imaging for treatment planning, patient positioning, and adaptive target tracking. However, radiographic imaging has the disadvantage of delivering a not-inconsequential concomitant dose to the patient, while CT imaging has imperfect soft tissue discrimination. These two concerns have prompted the development of external-beam delivery systems integrated with magnetic resonance imaging for intrafraction target alignment. The complex interaction of an MR imaging system with an external-beam delivery system creates a tradeoff problem in choosing and integrating the MR imaging technology and the external-beam technology. Lagendijk, Raaymakers, and Van Vulpen present the development of a high-field MR imaging system with a linear accelerator in Chapter 11, while Low describes the integration of a low-field

MR system with a movable array of Co-60 sources in Chapter 12. Both developers have devoted special attention to the problem of using MR images for both treatment planning and intrafraction real-time tumor targeting and tracking.

By necessity, motion adaptation puts beam alignment with the tumor into the hands of an automatic control system that allows only minimal human supervision and intervention during the treatment delivery. This level of automation requires exceptionally strict and robust error detection and correction methods to ensure that the dose is always delivered according to the treatment plan. Tracking errors, loss of target position, and improper beam adaptation responses must be detected, corrected, or stopped before any harm can be done. Some of the techniques currently in use to analyze reliability and manage the quality assurance of the delivery process are discussed by Sharp in Chapter 13.

Medical therapy with external-beams of radiation began as a two-dimensional technology in a three-dimensional world. The compromises of 2D spatial localization translated to large uncertainties in dose delivery. With the advent of computed tomography it became possible to design and deliver treatment plans in three spatial dimensions, which gave birth to 3D conformal and intensity-modulated radiation therapy and increased the accuracy and precision of dose delivery by many fold. However, in all but a limited number of scenarios, movement introduces the fourth dimension of time to the treatment problem. Motion-adaptive radiation therapy represents a truly four-dimensional solution to an inherently four-dimensional problem.

1

Real-Time Tumor Localization

1.1 Introduction 1
1.2 Tumor Localization Systems...1
    Radiographic Imaging • Electromagnetic Transponder • Fast MRI • Ultrasound
    • Positron Emission • Respiratory Monitoring Devices
1.3 Tumor Localization Methods 6
    Direct Methods • Indirect Methods • Hybrid Methods
1.4 Summary..9
References..9

Ruijiang Li

Laura I. Cerviño

Steve B. Jiang

1.1 Introduction

A critical step for motion-adaptive radiotherapy is the precise localization of the tumor in real time. An ideal localization system for radiotherapy should possess certain desirable properties (Dieterich et al. 2008). First, it has to be sufficiently accurate to capture real-time tumor motion and deformation caused by any physiological processes, such as respiration. Second, it should be also capable of imaging and localizing the surrounding structures in addition to the tumor. Last but not least, it should be noninvasive and present only minimal additional risk to the patient. Such an ideal localization system does not currently exist in clinics. However, there are several localization systems in current use or under active development that meet at least part of the above requirements for an ideal localization system. In the following sections, first we review these localization systems in detail. These cover a wide range of modern medical imaging modalities, including radiography, electronic portal imaging device (EPID), magnetic resonance imaging (MRI), ultrasound, radio frequency electromagnetic (EM) wave, optical, and positron emission, etc. Some are already commercially available and have been extensively used in clinics; some are under active development, while still others are research topics. Then we discuss different localization methods, given the hardware settings in different localization systems. Finally, we summarize and conclude our discussion.

1.2 Tumor Localization Systems

1.2.1 Radiographic Imaging

There are two categories of kilovoltage (kV) radiographic imaging systems for tumor tracking: room mounted and gantry mounted. Examples of commercially available room-mounted systems are the Mitsubishi/Hokkaido real-time tumor-tracking (RTRT) system, CyberKnife® system, and ExacTrac 6D system. Gantry-mounted systems include Elekta Synergy, Varian on-board imager (OBI), EPID, and integrated radiotherapy imaging system (IRIS) systems.

The RTRT system was developed jointly by Hokkaido University Hospital and Mitsubishi Electronics (Tokyo, Japan) (Shirato et al. 2000b). As shown in Figure 1.1a, the RTRT imaging system consists of four sets of diagnostic x-ray camera systems, each consisting of an x-ray tube mounted under the floor, a 9-in. image intensifier mounted on the ceiling, and a high-voltage x-ray generator. The four x-ray tubes are placed at right caudal, right cranial, left caudal, and left cranial positions with respect to the patient's couch at a distance of 280 cm from the isocenter. The image intensifiers are mounted on the ceiling, opposite to the x-ray tubes, at a distance of 180 cm from the isocenter, with beam central axes intersecting at the isocenter. At a given time during treatment of the patient, depending on the linac gantry angle, only two out of the four x-ray systems are enabled to provide a pair of orthogonal fluoroscopic images. To reduce the scatter radiation from the therapeutic beam to the imagers, the x-ray units and the linac are interleaved, that is, the megavoltage (MV) treatment beam is gated off when the kV x-ray units are pulsed. Although the RTRT system is capable of localizing the target in real time, it is only used for respiratory gating purposes.

CyberKnife is an image-guided radiosurgery system developed by Accuracy (Sunnyvale, CA). It consists of a 6-MV linac mounted on a robotic arm (Figure 1.1b). The robotic arm can point the beam anywhere in space with 6 degrees of freedom, without being constrained to a conventional isocenter (Adler et al. 1999). The CyberKnife system includes two x-ray tubes mounted on the ceiling and two amorphous silicon (aSi) flat panel imagers mounted at each side of the treatment couch (Ozhasoglu et al. 2008). The x-ray imaging system is used for initial alignment and periodic intrafraction tracking of the target position (at intervals of a few seconds to a few minutes). CyberKnife has been upgraded with a real-time respiratory localization and compensation system called Synchrony, which adds an infrared camera system for real-time monitoring of respiratory movement of the chest and the abdomen. Together with the x-ray cameras, this permits continuous compensation of respiratory movement. The Synchrony system is described in more detail in Chapter 4.

The ExacTrac 6D system developed by BrainLAB AG (Feldkirchen, Germany) is an integration of two subsystems (Figure 1.1c): (1) an infrared-based optical positioning system (ExacTrac), and (2) a radiographic kV x-ray imaging system

(X-Ray 6D) (Jin et al. 2008). The x-ray component consists of two floor-mounted kV x-ray tubes projecting obliquely into two corresponding aSi flat panel detectors mounted on the ceiling. The x-rays project in oblique directions relative to the patients, and the source isocenter and source detector distances are relatively large (224 and 362 cm, respectively).

Room-mounted x-ray imaging systems are particularly suitable for real-time internal fiducial marker localization during treatment. The x-ray sources and imagers are fixed on either floor or ceiling to provide high mechanical precision once calibrated. The imagers are far away from the patient so that the degradation of image quality by scattered MV photons is minimized. The downside of the large imager–patient distance is the

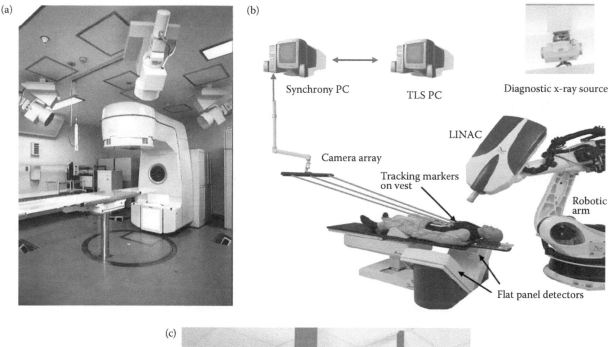

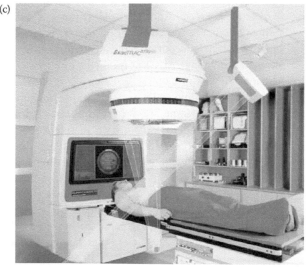

FIGURE 1.1 (a) Photograph of the Mitsubishi/Hokkaido Real-time tumor-tracking (RTRT) system; (b) Photograph of the CyberKnife Robotic Radiosurgery System with the Synchrony Respiratory Tracking System. (Courtesy of Ozhasoglu et al. 2008.); (c) Photograph of the BrainLAB ExacTrac 6D system showing the configurations of the x-ray imaging devices. (Courtesy of Jin et al. 2008.)

smaller field of view (FOV) and the lower imaging efficiency (which leads to higher imaging dose). One distinctive feature of room-mounted systems is the oblique projection angle, which makes human interpretation of the images difficult.

The Synergy system was developed by Elekta Inc. (Stockholm, Sweden) in collaboration with William Beaumont Hospital. The system consists of an x-ray tube and an aSi flat panel imager mounted on the linac gantry, orthogonal to the therapy beam (Figure 1.2a). A similar system called on-board imager, or OBI system (Figure 1.2b), was developed by Varian Medical Systems Inc. (Palo Alto, CA). Both Elekta Synergy and Varian OBI systems have only one imager. In general, two simultaneous projection images at oblique angles are required to locate a tumor in three dimensions (3D) unless the tumor follows essentially the same trajectory at all times. Unfortunately, tumors often follow complex 3D trajectories and sometimes exhibit hysteresis (Seppenwoolde et al. 2002). The limitations of a single kV imager for 3D tracking can be partially overcome by incorporating a trajectory estimation algorithm (Poulsen et al. 2008). IRIS system was developed by Massachusetts General Hospital (Berbeco et al. 2004a). As shown in Figure 1.2c, the system consists of two gantry-mounted kV x-ray tubes and two flat panel aSi imagers. The central axes of the two

kV x-ray beams are orthogonal to each other, 45° from the MV beam central axis, and intersecting with each other at the linac isocenter. The system was uniquely designed to integrate three main imaging functions (simultaneous orthogonal radiographs, cone-beam computerized tomography [CBCT], and real-time tumor localization) into a therapy system. Both x-ray sources are 100 cm away from the isocenter, while the imagers are at 162 cm distance from the isocenter. Fluoroscopic images can be acquired at a rate of 15 frames per second.

Gantry-mounted systems can be used to acquire radiographs with a large FOV at conventional beam angles (anterior, posterior, and lateral), as well as CBCT images. The major weakness of gantry-mounted systems is the suboptimal mechanical precision (e.g., gantry sagging) and the scatter radiation from the patient to the imagers.

Besides the above localization systems, which use kV x-ray sources within the treatment room, there is a different type of imaging system that uses MV treatment beam as its source. EPID has become a standard equipment on most modern linacs. In the cine mode of EPID, serial images can be acquired during treatment with minimal effort. There are several benefits of imaging during treatment with an EPID compared with a kV x-ray source: (1) No additional imaging dose to the patient is required.

(a)

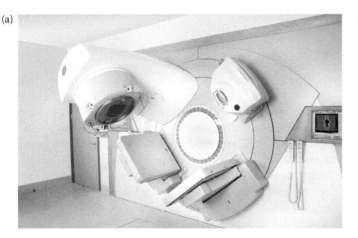

(b)

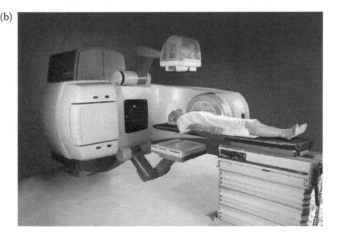

(c)

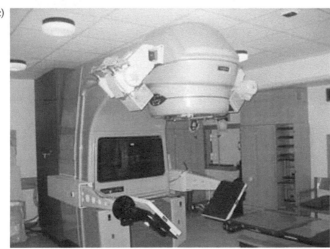

FIGURE 1.2 (a) Photograph of the Elekta Synergy system; (b) Photograph of the Varian OBI System; (c) Photograph of the IRIS system.

In this sense, the images from EPID are free; and (2) the images are acquired in the beam's eye view, which shows what is actually being irradiated. The main limitation of EPID is image quality. The contrast of soft tissues in MV images is inherently less than that in kV images because of increased Compton effects. This can be alleviated by implanting radio-opaque markers in the patient. Tumor localization using EPID is still an active research topic, and no clinical system is available yet (Balter et al. 2005; Terashima et al. 2005; Raaymakers et al. 2009).

1.2.2 Electromagnetic Transponder

Besides the use of radiographic imaging, technologies that use nonionizing EM fields are also available for real-time tumor localization. The Calypso 4D Localization System (Calypso Medical Technologies, Seattle, WA) is an example (Balter et al. 2005). The Calypso system comprises permanently implanted, miniature wireless EM transponders and a means for objective and continuous localization in 3D space. Essentially, these transponders are passive RLC resonant circuits encapsulated in a biocompatible glass capsule. The transponders are temporarily excited and induce resonant response signals. Each excited transponder briefly emits a magnetic field at a unique frequency. Sensors measure the emitted magnetic field strength to determine the coordinates. The size of the transponders is comparable to that of grains of rice (diameter 1.8 mm, length 8 mm). Because of their miniature dimensions, the implantation procedure is similar to standard seed placement or biopsy procedures in the prostate. Typically, a physician implants three transponders in the patient's body, usually directly into or in close proximity to the target. Currently, the prostate and the prostatic bed are the only implantation sites permitted by the U.S. Food and Drug Administration. The present size of the transponders, which requires a 14-gauge needle for implantation, limits their use in the lung.

1.2.3 Fast MRI

Fast MRI for real-time tumor localization is an emerging technology in radiotherapy that uses nonionizing EM fields. MRI yields better soft-tissue contrast than radiographic imaging. On-line localization of soft-tissue structures with MRI has been shown to be feasible and can be achieved at a subsecond time scale (Terashima et al. 2005). The major technical challenge is to integrate the MRI scanner with the linac and allow simultaneous MRI and irradiation without interfering with each other.

Recently, a prototype MRI/linac has been constructed by University Medical Center Utrecht in the Netherlands together with Elekta and Philips Research (Raaymakers et al. 2009). The prototype is a modified 6 MV Elekta linac, next to a modified 1.5 T Philips Achieva MRI system (Figure 1.3a). The two systems operate independently of each other and no degradation of the performance of either system was found in initial testing. This system is described in detail in Chapter 11. Another MRI/linac prototype has also been built at Cross Cancer Institute, Canada (Fallone et al. 2009), which consists of a 6 MV linac mounted onto the open end of a biplanar 0.2 T permanent MR system (Figure 1.3b). In addition to integration of MRI system with linac, MRI can be integrated with gamma ray radiotherapy. The Renaissance system under development by Dempsey et al. at ViewRay (Oakwood Village, OH) combines MRI with gamma ray radiotherapy, using three ⁶⁰Co sources with multileaf collimators (Figure 1.3c). This system is described in Chapter 12. Kron et al. (2006), at the Peter MacCallum Cancer Center in Australia, are also working on a cobalt-based treatment system (Figure 1.3d), integrating MR guidance with helical tomotherapy delivery.

Integration of MRI and radiotherapy system is a technology at its infancy. There are still many open questions that need to be investigated, for example, geometric accuracy of the MR images with millimeter precision, calibration of the coupling of coordinate systems between the MRI and radiotherapy system, and radiation dosimetry in the presence of a magnetic field (Lagendijk et al. 2008). Nonetheless, because of its superior soft-tissue contrast compared with conventional radiographic imaging, MRI will find its applications in all aspects of radiotherapy, treatment simulation, setup, and delivery, once the integration of MRI and radiotherapy system becomes a mature technology.

1.2.4 Ultrasound

Real-time 3D ultrasound is another alternative imaging modality for tumor localization that uses nonionizing radiation (Meeks et al. 2003). Ultrasonography is based on a pulse echo method in which a transducer emits brief ultrasound pulses and detects those that are reflected back when the sound wave meets a medium of a different density or compressibility. Feasibility of real-time target localization using ultrasound during radiotherapy treatment has been demonstrated in a phantom study (Hsu et al. 2005). It was found that the periodic noise in ultrasound images due to pulsing of the linac did not significantly degrade the accuracy of target localization and the presence of the transducer at the surface of the phantom presented minimal change to the dose distribution. Compared with many other imaging modalities, ultrasound has the additional benefits of being noninvasive, nonionizing, and economical. However, real-time tumor localization using ultrasound is still an active research topic, and no clinical implementation has been reported to date.

1.2.5 Positron Emission

A novel real-time tumor localization system that uses positron emission has recently been proposed and tested (Xu et al. 2006; Churchill et al. 2009). The system utilizes implanted positron emission markers that contain low-activity isotopes with half-lives comparable to the duration of the radiation therapy (few days to few weeks). By detecting annihilation gamma rays using two orthogonally placed position-sensitive detectors, multiple positron emission markers can be localized in real time.

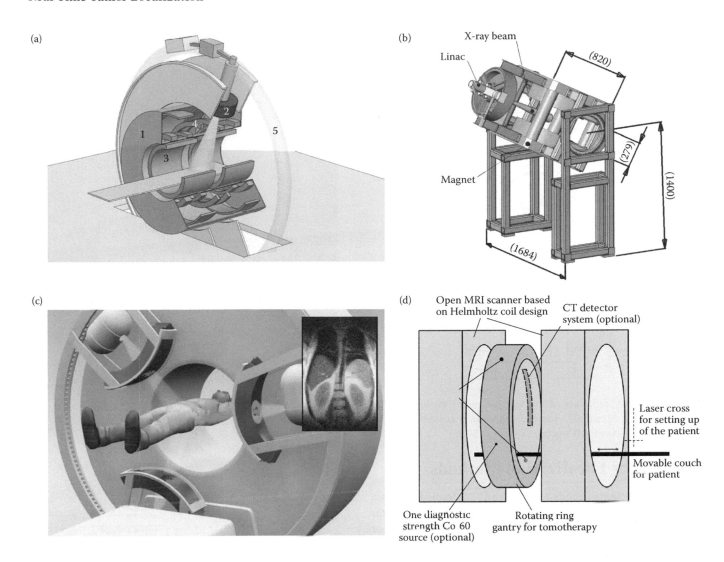

FIGURE 1.3 (a) The MRI/linac system prototype. (Courtesy of Raaymakers et al. 2009.); (b) The MRI/linac system prototype. (Courtesy of Fallone et al. 2009.); (c) Conceptual image of the system being designed by ViewRay. (Courtesy of James Dempsey.); (d) Conceptual image of MRI combined with cobalt-based helical tomotherapy delivery system. (Courtesy of Kron et al. 2006.)

Figure 1.4 depicts four detector modules mounted on the linac gantry, 50 cm from the isocenter. The imaging skin dose to the patient is much less compared with that in kV radiographic imaging, while a radiation dose as large as 18 Gy from the positron emission markers can be absorbed locally near the markers. This may have to be considered during planning for treatment. As of 2011, the system is still under development, and no clinical implementation has been reported so far.

1.2.6 Respiratory Monitoring Devices

Respiration is one of the major causes of tumor motion in the thorax and upper abdomen. Respiration can be monitored in different ways, which has led to the development of different systems. Early methods explored in the mid 1990s by Kubo and his colleagues include the use of thermistors, thermocouples, strain gauges, and pneumotachograph (Kubo and Hill 1996). They also developed a respiratory monitoring system that tracks infrared reflective markers placed on the patient's abdomen surface using a video camera, jointly with Varian Medical Systems Inc. (Kubo et al. 2000). It was later commercialized by Varian and was named real-time position management (RPM) system (Figure 1.5a). The RPM system has been extensively implemented and investigated clinically at a number of centers. Another commonly used device to monitor respiration is the spirometer (Wong et al. 1999; Zhang et al. 2003), which measures the time-integrated airflow and provides the lung volume information from a baseline (e.g., end of exhalation). Siemens Medical Systems (Concord, CA) has a monitoring interface that receives respiratory signals from a pressure cell on a belt around the patient that senses pressure changes as the patient breathes. More recently, 3D surface imaging systems that use video cameras are able to obtain surface images and track and monitor the patient's skin surface in real time. A commercially available 3D surface imaging system (Figure 1.5b) comes from VisionRT Ltd., London (Bert et al. 2005).

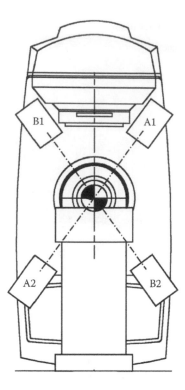

FIGURE 1.4 A depiction of the four-positron emission detector modules, A1, A2, B1, and B2, mounted on the linear accelerator gantry 50 cm from the isocenter. (Courtesy of Xu et al. 2006.)

1.3 Tumor Localization Methods

Tumor localization methods can be broadly classified into two groups: direct localization and indirect localization. Direct methods detect and localize the tumor or an object of interest within or near the tumor. Indirect methods detect and localize a surrogate and infer the spatial information of the tumor from the surrogate. In general, direct methods are more accurate, while indirect methods are noninvasive and have excellent temporal resolution. Combining direct and indirect methods leads to hybrid localization methods.

1.3.1 Direct Methods

Because of the vastly different imaging modalities used in different localization systems, some direct methods require implanted fiducial markers to detect and localize the tumor, while others do not. EM transponder and positron emission–based localization require implanted markers, and radiographic imaging may or may not require implanted markers depending on the tumor sites and image quality. MR- and ultrasound-based localizations do not require implanted markers for localization because of their higher soft-tissue contrast. Furthermore, they have the potential to provide real-time volumetric information of the tumor and other structures. Marker implantation is an invasive procedure, and it introduces additional risks to the patient. However, it is, in general, more accurate than the markerless counterpart.

Table 1.1 compares several marker-based direct localization methods. In terms of accuracy, all marker-based direct

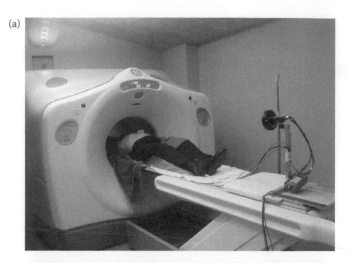

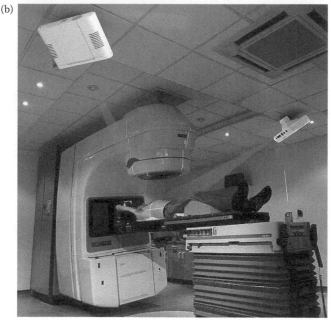

FIGURE 1.5 (a) The Varian RPM system; (b) The AlignRT system from VisionRT Ltd.

localization methods achieve a submillimeter localization accuracy. Among these, kV fluoroscopy has a relatively large FOV, which can potentially lead to localization of body structures other than the tumor. The FOV for other two localization modalities (EM transponder and Positron emission) is essentially restricted to the tumor. Marker type, size, tumor sites, and other medical conditions have to be taken into careful consideration before markers are implanted.

The most prevalent tumor localization system is the kV fluoroscopy with implanted markers, which is commercially available in the Mitsubishi/Hokkaido RTRT system. It has been extensively used for the treatment of lung, liver, prostate, and other tumors (Shirato et al. 2000; Seppenwoolde et al. 2002; Shirato et al. 1999; Shimizu et al. 2000; Shirato 2000; Shirato et al. 2003). In kV fluoroscopy localization, fiducial markers made of high-Z materials are implanted inside or near the tumor to help

TABLE 1.1 A Comparison of Marker-Based Direct Localization Methods

	Localization Modalities		
	kV Fluoroscopy	EM Transponder	Positron Emission
Marker type	Gold	Wireless EM transponders	Radioisotopes with short half-life
Marker size	1.5–2.0 mm diameter sphere	1.8 mm diameter and 8 mm long cylinder	0.8 mm diameter and 2–4 mm length
Localization accuracy	0.5 mm	0.2 mm	0.55 mm
Patient dose	0.3–1.2 Gy skin dose per hour	0.0 Gy	0.5–18 Gy lifetime dose to 4–12 cm^3 of tissue
Availability	Commercially available	Commercially available for prostate	Under development

Source: Xu, T., et al., *Med. Phys.*, 33, 2598–609, 2006.

localize the tumor position in real time. Spherical or cylindrical gold markers are often used for this purpose (Murphy et al. 2000; Mah et al. 2000). Markers can be implanted either percutaneously or endoscopically, depending on the tumor location and other medical considerations. The high radio-opacity of the markers makes them readily detectable in fluoroscopic images. Marker positions can be calculated through a fast and simple triangulation process, and markers are tracked at the video frame rate (30 Hz), thereby facilitating real-time localization.

MV radiographic imaging using EPID with implanted markers for real-time tumor localization is currently under study (Park et al. 2009; Berbeco et al. 2007). An automatic algorithm based on image processing techniques has been developed to extract the markers in EPID images and reconstruct 3D marker positions, which yields submillimeter localization accuracy (Park et al. 2009). Research on simultaneous kV and MV imaging for real-time localization of implanted fiducial markers is also underway (Wiersma et al. 2009; Liu et al. 2008; Mao et al. 2008a; Mao et al. 2008b; Wiersma et al. 2008; Cho et al. 2009). Localization accuracy below 1 mm in all the three spatial dimensions and frequency up to 10 Hz (limited by EPID acquisition speed) have been reported (Mao et al. 2008a). A potential problem using the MV beam for real-time localization is that the fiducial markers may be partially or completely blocked by the multileaf collimator at certain segments. This can be partially resolved by building a moving correlation model using both the present and past positions of the marker in the kV image to compute the position of the marker on the MV imager (Wiersma et al. 2009). Further research is warranted in this respect.

EM transponder-based localization is also commercially available (e.g., Calypso 4D Localization system). However, due to its relatively large marker (8 mm in length), implantation is currently limited to tumors in prostate (Balter et al. 2005; Kupelian et al. 2007; Willoughby et al. 2006). Localization can be done at a frequency of 10 Hz.

The markers used in positron emission-based localization have sizes comparable to those used in kV fluoroscopy (Xu et al. 2006). However, the topic is still under active research, and no clinical implementation is currently available.

Implanting fiducial markers percutaneously is an invasive procedure with potential risks of infection. Many clinicians are reluctant to perform this procedure for lung cancer treatment because puncturing the chest wall may cause pneumothorax (Arslan et al. 2002; Geraghty et al. 2003). Alternatively, one can place markers in the lung bronchoscopically; however, this has access limitations to different sites within the lung.

The effectiveness of using fiducial markers for tumor localization relies on the stability of the relationship between the markers and tumor. This relationship may change during the treatment course due to changes in tumor geometry. Additionally, markers may migrate between treatment simulation and delivery. For these reasons, three to four markers are often implanted, and marker migration can be detected by monitoring the intermarker spacing.

Owing to the risk of pneumothorax, percutaneous implantation of fiducial markers should be avoided whenever bronchoscopic placement is possible or the lung tumor can be visualized directly. Direct localization in fluoroscopy is extremely difficult, if not impossible, for tumors in the abdomen. In the case of lung tumors, however, the density difference between the tumor mass and normal lung tissue may be large enough to provide a good visualization in radiographic images. Early-stage lung cancer patients may have tumors of sufficient contrast with the surrounding lung tissues, and those patients may benefit from extracranial radiosurgery based on precise tumor localization. Berbeco et al. (Berbeco et al. 2004b) found that direct detection of a lung tumor mass in kV x-ray images is possible if the tumor mass is small and defined well and has a high-contrast edge. Subsequent work on markerless tumor localization in kV fluoroscopy has employed sophisticated machine learning and computer vision techniques (e.g., neural network, support vector machine, optical flow, and active shape models) (Cui et al. 2007; Xu et al. 2008; Lin et al. 2009; Xu et al. 2009). Localization can be performed at 15 Hz (limited by fluoroscopy acquisition speed), and accuracy in the order of 1–2 mm has been reported. A major hurdle for these methods is that they require training fluoroscopic data to be acquired prior to each treatment, in which a clinician must define tumor positions in each fluoroscopic image. In principle, digitally reconstructed radiograph templates made from Four-dimensional computed tomography (4DCT) can be used to simulate training fluoroscopic images, where the tumor position has already been marked. In addition, fluoroscopic images from only anterior-posterior direction (AP) direction have been analyzed, where tumor has a higher contrast. Preliminary work on markerless tumor localization in EPID images has also been done (Arimura et al. 2009). Localization error is around 1.5 mm, but the processing time is quite long (around 1.8 seconds per frame). To be clinically useful, further investigation is needed on markerless tumor localization.

1.3.2 Indirect Methods

When direct localization of the tumor is difficult (e.g., in kV fluoroscopy without implanted markers), it may be beneficial to use indirect methods, which infer the spatial information of the tumor from some kind of surrogates that are easily localized. For respiration-induced tumor motion, localization can be done using the so-called respiratory surrogates. Surrogates can be further categorized into two classes: internal surrogates (diaphragm, carina, etc.) and external surrogates (optical skin markers, body surface, tidal volume, etc.). If the tumor is close to the internal surrogates, the internal surrogates will provide more accurate localization than external ones. On the other hand, external surrogates are easier to localize and pose minimal risk to the patient, but they may introduce more sources of errors compared with internal surrogates.

Internal surrogates are anatomical surrogates present inside the patient's body. Intuitively, the closer the surrogate is to the tumor, the more accurate the localization will be, provided the surrogate can be easily localized. For instance, diaphragm may be a good surrogate for tumors in the lower lung, liver, and pancreas; carina may be good for localizing tumors in the upper lung. Dawson et al. found that diaphragm motion correlates within 2 mm with microcoils implanted near hepatic tumors (Dawson et al. 2001). Correlation of lung tumor motion with the diaphragm or chest wall shows a much greater variation (Hoisak et al. 2004). Cervino et al. analyzed the correlation between diaphragm motion and the superior–inferior (SI) lung tumor motion in fluoroscopic image sequences from ten lung cancer patients (Cervino et al. 2009). Average correlation of 0.98 can be achieved with a linear model accounting for phase delay, and an average localization error of 0.8 mm was reported. However, in one patient studied, the correlation is much weaker than that in other patients. Van der Weide et al. (van der Weide et al. 2008; Spoelstra et al. 2009) preferred using carina as a respiratory surrogate, which gave an average localization error of 2–3 mm on repeat 4DCT scans from 23 patients. However, in the majority of the patients (19 out of 23), tumors are located in the upper lobe of the lung (Spoelstra et al. 2009), which might have favored the use of carina. Unlike diaphragm, which is clearly discernible in kV fluoroscopy, the carina may not be easily detected. The applicability of each surrogate for real-time tumor localization needs to be further investigated and should be assessed on individual basis (Mah et al. 2000; Cervino et al. 2009; Stevens et al. 2001).

External surrogates are patient's skin surface or artificial objects in contact with the patient's body. The most commonly used external surrogate is the AP motion of the abdominal surface (Wu et al. 2008; Jiang 2006; Kanoulas et al. 2007; Ruan

et al. 2008). Fluoroscopic studies with the RPM system have demonstrated high short-term (1 minute) correlation between abdominal surface motion and diaphragm motion in most cases (Vedam et al. 2003; Mageras et al. 2001). A study by Wagman et al. (Wagman et al. 2003) found good reproducibility in abdominal organ positions with prospectively triggered CT at end expiration using the RPM signal, with average organ displacement of 0.2 cm in the SI direction in repeat CT scans at the same session. Studies on the lung tumor localization from skin motion reported an average error of 5.4 mm (Ahn et al. 2004). Tidal volume measurements using spirometers seem to be a better respiratory surrogate, where average localization error of 2–4 mm has been reported (Zhang et al. 2003).

A key issue for indirect localization using surrogates is accuracy because it relies on the assumption that there is a good correlation between tumor motion and respiratory surrogates and that the correlation is stable over the entire treatment course. Studies have shown that the correlation between tumor motion and abdominal surface motion depends on individual patients and can change both within and between treatment fractions (Hoisak et al. 2004; Bruce 1996; Hoisak et al. 2006; Tsunashima et al. 2004). Specifically, internal–external correlation can be disturbed by transient changes in breathing (Ozhasoglu and Murphy 2002). Drifts of the surrogates may occur, caused by patient movement, particularly if the motion amplitude is small, such as while monitoring anterior chest wall motion (Chen et al. 2001). Therefore, it is important to coach the patients and evaluate their ability to breathe in a regular and reproducible way during the treatment (Jiang 2006; Kini et al. 2003; Neicu et al. 2006). Even if respiratory surrogates correlate well with the tumor motion within a single treatment session, the relationship between surrogates and internal tumor motion may still change between sessions, which can adversely affect organ reproducibility. Factors that can affect the diaphragm respiratory signal relationship between sessions include changes in patient's respiration pattern (such as the relative amount of chest versus abdominal displacement during breathing), changes in abdominal pressure caused by stomach filling, and changes in hepatic tumor shape and size.

For the above reasons, tumor localization with surrogates should be used with great caution. In order to infer the amount of tumor motion for a given surrogate displacement, the relative magnitudes of tumor and surrogate motion should be measured during simulation and each treatment fraction. The correlation between surrogate and tumor motion should be established before each treatment fraction and checked and updated during the treatment session at a frequency depending on the tumor site and even the individual patient (Chen et al. 2001; Schweikard et al. 2000).

1.3.3 Hybrid Methods

There is a third type of localization method called hybrid method, which combines some of the benefits of both direct and indirect methods. It is more accurate than indirect methods and poses less risk to the patients (e.g., radiation dose in kV fluoroscopy) than direct methods. Hybrid methods continuously monitor external respiratory signals to localize internal tumor motion and acquire episodic radiographic images to update the internal–external motion model, based on the assumption that external surrogates can accurately predict the internal tumor motion during the time interval between radiographic image acquisitions (Ozhasoglu and Murphy 2002; Chen et al. 2001; Neicu et al. 2006; Schweikard et al. 2000; Murphy 2004; Murphy et al. 2002; Sharp et al. 2004). Some commercial integrated localization systems use hybrid methods. For instance in CyberKnife Synchrony system, besides two pairs of x-ray sources and imagers, a camera array with three charge-coupled device cameras mounted on the ceiling is used to track the light-emitting diode markers attached to a patient's vest at a rate of 25–40 Hz (Ozhasoglu et al. 2008). At the beginning of treatment, two orthogonal x-ray images are taken at least eight times at different phases of the respiratory cycle, and the positions of the internal fiducial markers are calculated. Then a regression model is built between the external and internal markers and is used to predict the tumor position based on the position of external optical markers. Whenever x-ray images are taken, positions of internal fiducial markers are recalculated and used to update or rebuild the regression model. The radiographic imaging frequency is a compromise between localization accuracy and radiation dose to the patient. The CyberKnife hybrid correlation technique is described in detail in Chapters 4 and 7.

1.4 Summary

We have reviewed different real-time tumor localization systems and different localization methods (direct, indirect, and hybrid) for radiation therapy. One of the major limitations with current clinical systems is that they can only track translational changes of a single point (e.g., center of the tumor). However, deformational changes in tumor are known to be present between fractions and even within fractions. In addition, there may be translational as well as deformational changes in the surrounding normal tissues too. Real-time volumetric imaging techniques such as MRI and ultrasound have great potential to capture these changes. As science and technology advance, current tumor localization systems will evolve toward the ideal localization system for radiotherapy.

References

Adler, J.R. Jr., et al., Image-guided robotic radiosurgery. *Neurosurgery*, 44(6), 1299–306, discussion 1306–7, 1999.

Ahn, S., et al., A feasibility study on the prediction of tumour location in the lung from skin motion. *Br J Radiol*, 77(919), 588–96, 2004.

Arimura, H., et al., Computerized method for estimation of the location of a lung tumor on EPID cine images without implanted markers in stereotactic body radiotherapy. *Phys Med Biol*, 54(3), 665–77, 2009.

Arslan, S., et al., CT- guided transthoracic fine needle aspiration of pulmonary lesions: accuracy and complications in 294 patients. *Med Sci Monit*, 8(7), CR493–97, 2002.

Balter, J.M., et al., Accuracy of a wireless localization system for radiotherapy. *Int J Radiat Oncol Biol Phys*, 61(3), 933–37, 2005.

Berbeco, R.I., et al., Integrated radiotherapy imaging system (IRIS): design considerations of tumour tracking with linac gantry-mounted diagnostic x-ray systems with flat-panel detectors. *Phys Med Biol*, 49(2), 243–55, 2004a.

Berbeco, R.I., et al. "Tumor tracking in the absence of radiopaque markers," in *The 14th International Conference on the Use of Computers in Radiation Therapy*, Seoul, Korea, 2004b.

Berbeco, R.I., et al., Clinical feasibility of using an EPID in CINE mode for image-guided verification of stereotactic body radiotherapy. *Int J Radiat Oncol Biol Phys*, 69(1), 258–66, 2007.

Bert, C., et al., A phantom evaluation of a stereo-vision surface imaging system for radiotherapy patient setup. *Med Phys*, 32(9), 2753–62, 2005.

Bruce, E.N., Temporal variations in the pattern of breathing. *J Appl Physiol*, 80(4), 1079–87, 1996.

Cervino, L.I., et al., The diaphragm as an anatomic surrogate for lung tumor motion. *Phys Med Biol*, 54(11), 3529–41, 2009.

Chen, Q.S., et al., Fluoroscopic study of tumor motion due to breathing: facilitating precise radiation therapy for lung cancer patients. *Med Phys*, 28(9), 1850–56, 2001.

Cho, B, Poulsen, P.R, Sloutsky, A, Sawant, A, and Keall, P.J, First demonstration of combined kV/MV image-guided real-time multi-leaf collimator target tracking, *Int J Radiat Oncol Biol Phys*, 74 (3), 859–67, 2009.

Churchill, N.W., Chamberland, M., and Xu, T., Algorithm and simulation for real-time positron emission based tumor tracking using a linear fiducial marker. *Med Phys*, 36(5), 1576–86, 2009.

Cui, Y., et al., Multiple template-based fluoroscopic tracking of lung tumor mass without implanted fiducial markers. *Phys Med Biol*, 52(20), 6229–42, 2007.

Dawson, L.A., et al., The reproducibility of organ position using active breathing control (ABC) during liver radiotherapy. *Int J Radiat Oncol Biol Phys*, 51(5), 1410–21, 2001.

Dieterich, S., et al., Locating and targeting moving tumors with radiation beams. *Med Phys*, 35(12), 5684–94, 2008.

Fallone, B.G., et al., First MR images obtained during megavoltage photon irradiation from a prototype integrated linac-MR system. *Med Phys*, 36(6), 2084–88, 2009.

Geraghty, P.R., et al., CT-guided transthoracic needle aspiration biopsy of pulmonary nodules: needle size and pneumothorax rate. *Radiology*, 229(2), 475–81, 2003.

Hoisak, J.D., et al., Prediction of lung tumour position based on spirometry and on abdominal displacement: accuracy and reproducibility. *Radiother Oncol*, 78(3), 339–46, 2006.

Hoisak, J.D., et al., Correlation of lung tumor motion with external surrogate indicators of respiration. *Int J Radiat Oncol Biol Phys*, 60(4), 1298–306, 2004.

Hsu, A., et al., Feasibility of using ultrasound for real-time tracking during radiotherapy. *Med Phys*, 32(6), 1500–12, 2005.

Jiang, S.B., Technical aspects of image-guided respiration-gated radiation therapy. *Med Dosim*, 31(2), 141–51, 2006.

Jin, J.Y., et al., Use of the BrainLAB ExacTrac X-Ray 6D system in image-guided radiotherapy. *Med Dosim*, 33(2), 124–34, 2008.

Kanoulas, E., et al., Derivation of the tumor position from external respiratory surrogates with periodical updating of the internal/external correlation. *Phys Med Biol*, 52(17), 5443–56, 2007.

Kini, V.R., et al., Patient training in respiratory-gated radiotherapy. *Med Dosim*, 28(1), 7–11, 2003.

Kron, T., et al., Magnetic resonance imaging for adaptive cobalt tomotherapy: A proposal. *J Med Phys*, 31(4), 242–54, 2006.

Kubo, H.D. and B.C. Hill, Respiration gated radiotherapy treatment: a technical study. *Phys Med Biol*, 41(1), 83–91, 1996.

Kubo, H.D., et al., Breathing-synchronized radiotherapy program at the University of California Davis Cancer Center. *Med Phys*, 27(2), 346–53, 2000.

Kupelian, P., et al., Multi-institutional clinical experience with the Calypso System in localization and continuous, real-time monitoring of the prostate gland during external radiotherapy. *Int J Radiat Oncol Biol Phys*, 67(4), 1088–98, 2007.

Lagendijk, J.J., et al., MRI/linac integration. *Radiother Oncol*, 86(1), 25–29, 2008.

Lin, T., et al., Fluoroscopic tumor tracking for image-guided lung cancer radiotherapy. *Phys Med Biol*, 54(4), 981–92, 2009.

Liu, W., et al., Real-time 3D internal marker tracking during arc radiotherapy by the use of combined MV-kV imaging. *Phys Med Biol*, 53(24), 7197–213, 2008.

Mageras, G.S., et al., Fluoroscopic evaluation of diaphragmatic motion reduction with a respiratory gated radiotherapy system. *J Appl Clin Med Phys*, 2(4), 191–200, 2001.

Mah, D., et al., Technical aspects of the deep inspiration breath-hold technique in the treatment of thoracic cancer. *Int J Radiat Oncol Biol Phys*, 48(4), 1175–85, 2000.

Mao, W., et al., A fiducial detection algorithm for real-time image guided IMRT based on simultaneous MV and kV imaging. *Med Phys*, 35(8), 3554–64, 2008a.

Mao, W., R.D. Wiersma, and L. Xing, Fast internal marker tracking algorithm for onboard MV and kV imaging systems. *Med Phys*, 35(5), 1942–49, 2008b.

Meeks, S.L., et al., Ultrasound-guided extracranial radiosurgery: technique and application. *Int J Radiat Oncol Biol Phys*, 55(4), 1092–101, 2003.

Murphy, M.J., et al., Image-guided radiosurgery for the spine and pancreas. *Comput Aided Surg*, 5(4), 278–88, 2000.

Murphy, M.J., Tracking moving organs in real time. *Semin Radiat Oncol*, 14(1), 91–100, 2004.

Murphy, M.J., J. Jalden, and M. Isaksson. "Adaptive filtering to predict lung tumor breathing motion during image-guided radiation therapy," in *Proc. 16th Int. Conf. on Computer Assisted Radiology (CARS 2002)*. Paris, 2002.

Neicu, T., et al., Synchronized moving aperture radiation therapy (SMART): improvement of breathing pattern reproducibility using respiratory coaching. *Phys Med Biol*, 51(3), 617–36, 2006.

Ozhasoglu, C. and M.J. Murphy, Issues in respiratory motion compensation during external-beam radiotherapy. *Int J Radiat Oncol Biol Phys*, 52(5), 1389–99, 2002.

Ozhasoglu, C., et al., Synchrony—Cyberknife respiratory compensation technology. *Med Dosim*, 33(2), 117–23, 2008.

Park, S.J., et al., Automatic marker detection and 3D position reconstruction using cine EPID images for SBRT verification. *Med Phys*, 36(10), 4536–46, 2009.

Poulsen, P.R, Cho, B., Langen, K., Kupelian, P., and Keall, P.J., Three-dimensional prostate position estimation with a single x-ray imager utilizing spatial probability density. *Phys Med Biol*, 53(16), 4331–53, 2008.

Raaymakers, B.W., et al., Integrating a 1.5 T MRI scanner with a 6 MV accelerator: proof of concept. *Phys Med Biol*, 54(12), N229–37, 2009.

Ruan, D., et al., Inference of hysteretic respiratory tumor motion from external surrogates: a state augmentation approach. *Phys Med Biol*, 53(11), 2923–36, 2008.

Schweikard, A., et al., Robotic motion compensation for respiratory movement during radiosurgery. *Comput Aided Surg*, 5(4), 263–77, 2000.

Seppenwoolde, Y., et al., Precise and real-time measurement of 3D tumor motion in lung due to breathing and heartbeat, measured during radiotherapy. *Int J Radiat Oncol Biol Phys*, 53(4), 822–34, 2002.

Sharp, G.C., et al., Prediction of respiratory tumour motion for real-time image-guided radiotherapy. *Phys Med Biol*, 49(3), 425–40, 2004.

Shimizu, S., et al., Use of an implanted marker and real-time tracking of the marker for the positioning of prostate and bladder cancers. *Int J Radiat Oncol Biol Phys*, 48(5), 1591–97, 2000.

Shirato, H., et al., Real-time tumour-tracking radiotherapy. *Lancet*, 353(9161), 1331–32, 1999.

Shirato, H., et al., Four-dimensional treatment planning and fluoroscopic real-time tumor tracking radiotherapy for moving tumor. *Int J Radiat Oncol Biol Phys*, 48(2), 435–42, 2000a.

Shirato, H., et al., Physical aspects of a real-time tumor-tracking system for gated radiotherapy. *Int J Radiat Oncol Biol Phys*, 48(4), 1187–95, 2000b.

Shirato, H., et al., Feasibility of insertion/implantation of 2.0-mm-diameter gold internal fiducial markers for precise setup and real-time tumor tracking in radiotherapy. *Int J Radiat Oncol Biol Phys*, 56(1), 240–47, 2003.

Spoelstra, F.O., et al., An evaluation of two internal surrogates for determining the three-dimensional position of peripheral lung tumors. *Int J Radiat Oncol Biol Phys*, 74(2), 623–29, 2009.

Stevens, C.W., et al., Respiratory-driven lung tumor motion is independent of tumor size, tumor location, and pulmonary function. *Int J Radiat Oncol Biol Phys*, 51(1), 62–8, 2001.

Terashima, M., et al., High-resolution real-time spiral MRI for guiding vascular interventions in a rabbit model at 1.5 T. *J Magn Reson Imaging*, 22(5), 687–90, 2005.

Tsunashima, Y., et al., Correlation between the respiratory waveform measured using a respiratory sensor and 3D tumor motion in gated radiotherapy. *Int J Radiat Oncol Biol Phys*, 60(3), 951–58, 2004.

van der Weide, L., et al., Analysis of carina position as surrogate marker for delivering phase-gated radiotherapy. *Int J Radiat Oncol Biol Phys*, 71(4), 1111–117, 2008.

Vedam, S.S., et al., Quantifying the predictability of diaphragm motion during respiration with a noninvasive external marker. *Med Phys*, 30(4), 505–13, 2003.

Wagman, R., et al., Respiratory gating for liver tumors: use in dose escalation. *Int J Radiat Oncol Biol Phys*, 55(3), 659–68, 2003.

Wiersma, R.D., Mao, W., and Xing, L., Combined kV and MV imaging for real-time tracking of implanted fiducial markers. *Med Phys*, 35(4), 1191–98, 2008.

Wiersma, R.D., et al., Use of MV and kV imager correlation for maintaining continuous real-time 3D internal marker tracking during beam interruptions. *Phys Med Biol*, 54(1), 89–103, 2009.

Willoughby, T.R., et al., Target localization and real-time tracking using the Calypso 4D localization system in patients with localized prostate cancer. *Int J Radiat Oncol Biol Phys*, 65(2), 528–34, 2006.

Wong, J.W., et al., The use of active breathing control (ABC) to reduce margin for breathing motion. *Int J Radiat Oncol Biol Phys*, 44(4), 911–19, 1999.

Wu, H., et al., Gating based on internal/external signals with dynamic correlation updates. *Phys Med Biol*, 53(24), 7137–50, 2008.

Xu, T., et al., Real-time tumor tracking using implanted positron emission markers: concept and simulation study. *Med Phys*, 33(7), 2598–609, 2006.

Xu, Q., et al., A deformable lung tumor tracking method in fluoroscopic video using active shape models: a feasibility study. *Phys Med Biol*, 52(17), 5277–93, 2007.

Xu, Q., et al., Lung Tumor Tracking in Fluoroscopic Video Based on Optical Flow. *Med Phys*, 35(12), 5351–59, 2008.

Zhang, T., et al., Application of the spirometer in respiratory gated radiotherapy. *Med Phys*, 30(12), 3165–71, 2003.

Theoretical Aspects of Target Detection and Tracking

2.1   The Likelihood Ratio Test...13
2.2   Sensor Modeling 14
2.3   Implications of Detection Theory 15
2.4   Image-Based Tracking 15
2.5   Template Selection and Motion Enhancement 15
2.6   Matching Cost Functions...16
        Popular Matching Cost Functions • Robust Matching Cost Functions
2.7   Tracking and Prediction...18
        Single Object Tracking • Multiobject Tracking
References 20

Gregory C. Sharp

Rui Li

Nagarajan Kandasamy

Image-based trackers are a flexible and robust solution to the problem of finding an internal target. They can be used alone or in conjunction with surrogates. They are usually used as a point localization device but can also be used to find larger objects or curves. Images can be used to localize implanted objects and natural anatomical landmarks, and they can be processed to track multiple targets. The chief disadvantage of image-based trackers is that they are usually implemented using fluoroscopic imaging, which means an extra radiation burden to the patient under treatment. Management of fluoroscopic imaging dose can be accomplished by reducing either the imaging technique or the number of images. These dose management methods, however, make object localization and tracking more difficult. Robust and reliable detection in image-based tracking systems is therefore desirable to achieve accurate real-time localization.

This chapter introduces the basic theory of object detection and describes how it is implemented in commonly used image-based detection and tracking systems.

2.1 The Likelihood Ratio Test

Detection theory is concerned with making a binary decision based on the results of an experiment. Our concern is the detection of an object, such as a radio-opaque marker, within an image. We therefore form a hypothesis that the marker is located at position (x,y) and an opposite hypothesis (called the null hypothesis) that the marker is not located there.

H_1 The object is located at (x,y)
H_0 The object is not located at (x,y)

Given sensor information θ and a suitable model of the sensor, we can compute the likelihood of a hypothesis H being true given θ as follows:

$$L(\mathrm{H} \mid \theta) = P(\theta \mid \mathrm{H}). \qquad (2.1)$$

The term $P(\theta|\mathrm{H})$ is the probability of sensing θ given that hypothesis H is true. The form of this term depends upon the sensor model and the hypothesis.

The likelihoods for two different hypotheses can be compared to determine which of them is true. This is usually done using a measure called the "likelihood ratio," which is given as

$$\Lambda(\theta) = \frac{L(\mathrm{H}_0 \mid \theta)}{L(\mathrm{H}_1 \mid \theta)}. \qquad (2.2)$$

The likelihood ratio test compares a likelihood ratio with a predetermined threshold, and is used to decide between two hypotheses.

We illustrate the use of the likelihood ratio test in finding a radio-opaque marker in a fluoroscopic image. Consider three different views of the same marker under different imaging conditions, as shown in Figure 2.1. The marker is a gold cylinder of 0.8 mm diameter and 2 mm length suspended in 23 cm of solid water to mimic the imaging conditions of an adult liver. The imaging system comprises an x-ray tube operating in a pulsed fluoro mode at 125 kVp and 0.2 mAs per pulse and a fluoro-capable amorphous silicon flat panel detector mounted on the gantry of a linear accelerator. In the left panel, the 6 MV treatment beam is *off*, and we can see a clear image of the marker relative to

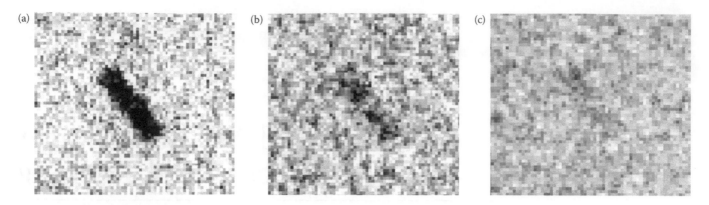

FIGURE 2.1 A radio-opaque marker seen under three different imaging conditions: (a) without MV scatter; (b) with scatter from 300 MU/min 10 × 10 MV beam; and (c) with scatter from 300 MU/min 20 × 20 MV beam. Detectability degrades when scattered radiation from a treatment device hits the imaging panel.

the background. In the center and right panels, the 6 MV treatment beam is *on* at 300 MU per minute with either a 10 × 10 or 20 × 20 field size.

To estimate the detectability of the marker, we model the image region as a collection of pixels that belong to either the foreground object (the marker) or the background. Both the foreground intensity P_f, and the background intensity P_b, are assumed to be uniform, and the image is corrupted by Gaussian noise. Thus,

$$P_f \sim \mathcal{N}(\mu_f, \sigma),$$
$$P_b \sim \mathcal{N}(\mu_b, \sigma), \qquad (2.3)$$

where μ_f and μ_b are the modeled intensity of the foreground and background, respectively, and σ is the standard deviation of the noise.

We now derive the formulas for $P(\theta|H_0)$ and $P(\theta|H_1)$ under the assumption that each pixel is an independent measurement. The image region θ is divided into foreground pixels x_f within the foreground area θ_f and background pixels x_b within the background area θ_b. The probability for sensing any given pixel depends on whether it is a foreground pixel or background pixel. These probabilities are given as

$$P_f(x_f) = \frac{1}{\sqrt{2\pi\sigma^2}} e^{-(x-\mu_f)^2/(2\sigma^2)} \quad \text{and}$$

$$P_b(x_b) = \frac{1}{\sqrt{2\pi\sigma^2}} e^{-(x-\mu_b)^2/(2\sigma^2)}. \qquad (2.4)$$

Therefore, probability for sensing the collection of pixels within region θ is given as

$$P(\theta|H_0) = \prod_{x_b \in \theta_b} P_b(x_b) \prod_{x_f \in \theta_f} P_b(x_f) \quad \text{and}$$

$$P(\theta|H_1) = \prod_{x_b \in \theta_b} P_b(x_b) \prod_{x_f \in \theta_f} P_f(x_f). \qquad (2.5)$$

When using these probabilities to compute the likelihood ratio test, we see that the pixels that comprise the background

area θ_b have no influence. This is because the background area will look the same whether or not the radio-opaque marker is present. Furthermore, because the test compares the likelihood ratio against a threshold, we can use the log likelihood measure instead. With these considerations in mind, an equivalent log likelihood ratio test can be formulated as

$$\Lambda(\theta) \sim \sum_{x_f \in \theta_f} (x_f - \mu_b)^2 - \sum_{x_f \in \theta_f} (x_f - \mu_f)^2. \qquad (2.6)$$

In this equation, $\Lambda(\theta)$ is composed of two terms: the sum of squared difference between the sensed values and the ideal background and foreground values. When the difference between these two terms is larger than the threshold, the object can be detected.

2.2 Sensor Modeling

We now turn our attention to modeling the fluoroscopic sensor. The model of Equation 2.3 has three terms: the intensity values of the foreground, the intensity values of the background, and the standard deviation of the noise. In principle, these quantities can be derived experimentally using phantom experiments.

To illustrate this idea, experiments have been performed on the integrated radiotherapy imaging system (cf. Chapter 1) using the experimental setup described in Section 2.1. The radio-opaque marker was placed in 23 cm of solid water and imaged at 125 kVp with varying mAs. The mean intensity values for the signal and background vary with imaging conditions, as shown in Figure 2.2 (left). We are primarily interested in the difference between signal and background, and from the figure, we confirm that signal increases linearly with increased mAs.

Image noise is modeled by two sources: the kV imaging system itself and the scatter from the MV source. These contributions are shown in the center and right of Figure 2.2. The kV imager includes an offset term due to electronic noise in the panel and a linear term, which increases with increasing mAs. The MV term varies with dose rate and field size.

Once the sensor has been modeled, as demonstrated in Figure 2.2, we can use the model to make design decisions. For example,

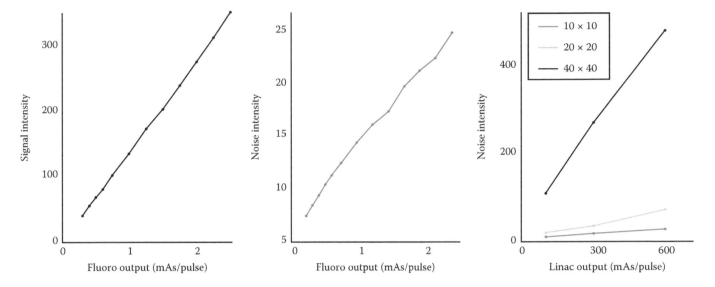

FIGURE 2.2 Signal and noise intensity as a function of x-ray generator and linac output for three different field sizes.

TABLE 2.1 Fluoroscopic Technique Needed for the System in Figure 2.2 as a Function of Dose Rate and Field Size

	100 MU/min	300 MU/min	600 MU/min
10 × 10	<0.2 mAs/pulse	<0.2 mAs/pulse	<0.2 mAs/pulse
20 × 20	<0.2 mAs/pulse	<0.2 mAs/pulse	0.3 mAs/pulse
40 × 40	0.4 mAs/pulse	1.2 mAs/pulse	2.0 mAs/pulse

we might want to use Equation 2.6 to achieve the objective of 99.99% correct detection by our implanted marker within a region of interest. The theoretical mAs required to achieve this goal are shown in Table 2.1.

2.3 Implications of Detection Theory

In the preceding example, we have derived the theoretical basis of detection using a mean-squared error tracker and have also determined the imaging technique required for guaranteed performance. These derivations are, however, quite limited in their application. First, our model considered only a foreground object in a uniform background. However, the reality is that implanted markers are not usually placed in regions of uniform background. The background will be irregular and unpredictable and may change over time. Explicit modeling of the background clutter is difficult and usually not worth the effort.

Nevertheless, detection theory yields an important insight which should not be lost. Specifically, only two factors contribute to the detection rate in an image-based tracker:

1. Signal-to-noise ratio
2. Size of the foreground object

We see this relationship clearly in Equation 2.6. Therefore, any effort to improve the detection rate should be concentrated on

these two items. Signal-to-noise ratio can be increased by increasing the mAs or by using a marker with higher atomic number. Increasing the size of the foreground object means using a larger marker or, perhaps, tracking a larger region of interest.

2.4 Image-Based Tracking

Image-based tracking is the process of locating one or more moving objects within an input video sequence. We limit our scope of discussion to 2D image-based tracking using a single input video sequence. There are two critical components in image-based tracking. The first component is object detection, which determines the appearance and location of the tracked objects (objects of interest). Unless the signal-to-noise ratio is very high, there will be multiple locations detected, some of which are false positives. We can view all these detections as hypotheses of possible object locations. The second component is temporal filtering and data association, which handles the dynamics of the objects of interest and evaluates different hypotheses. Depending on the application, these two components are combined and weighted differently.

Thus in the following sections, we first introduce object representations using templates and then discuss ways to enhance the appearance in tracking. Various cost functions used in object localization are described in Section 2.6. Then we discuss robust cost functions, which are less sensitive to noise, in Section 2.6.2. In Section 2.7, we briefly discuss some commonly used approaches for temporal filtering and prediction.

2.5 Template Selection and Motion Enhancement

If a template describing a specific object is available, object detection becomes a process of matching features between the template and the image sequences. Object detection is generally computationally expensive, and the accuracy of the detections depends heavily on the details and the degree of precision provided by the object template. The most commonly used template is either a manually predetermined template or a template parameterized by a function.

A predetermined template can be obtained from the input images (empirical template) or constructed manually (model-based template) as shown in lower right corner of Figure 2.3. The disadvantage of using an empirical template is that the cropped image region may contain too little object information and too much background. As a result, many false matches might be detected. A remedy to this problem is to use a model-based template as shown in the lower left corner of Figure 2.3. The example model-based template does not model background image information but contains strong signal strength for the object we wish to detect. Pixels of the model-based template are weighted in the range of 0–1 based on their distance from the object of interest. In Figure 2.3, all pixels inside the inner band carry weight of 1 and all pixels outside the outer band carry weight of 0. The pixels in between the two bands carry weights between 0 and 1, which are inversely proportional to their distances from the object of interest.

If the object(s) of interest and the background structures have distinctive motions, then we can make use of the motion information to remove the background structures and enhance the contrast between the objects and background. As a result, the objects of interest can be easily identified, and false matches can be eliminated in this motion enhancement step. The motion-enhanced images in Figure 2.4 are obtained by subtracting the mean image from the input image frames. The mean image is computed from a set of training image frames. The key assumption is that the background motion and object motion are distinct; hence, the subtraction of the mean image will enhance the area where motion differences are large. In the first row, we see that motion enhancement eliminates stationary background objects such as the vertebrae, and the only distinctive object is the radio-opaque marker. However, in the second row of Figure 2.4, the input image has a bright region, which is an air bubble near the marker. Because the

movement of the air bubble is similar to that of the implanted radio-opaque markers, mean image subtraction does not work effectively in this case. This calls for more sophisticated motion enhancement methods as discussed in other studies [SCI05, CFJ08].

In deformable template matching, a template is represented as a bitmap describing the characteristic contour or edges of an object shape. Thus, we can view this template as a prototype contour of the object of interest. During matching, a probabilistic transformation is applied to the prototype contour to deform the template to fit edge features in the input image. Such a transformation is described by a set of parameters, and the matching process aims to find the best parameters by optimizing an objective function to produce the best match between deformed template and image features. We refer interested readers to [JZL96, ZJDJ00, SL01] for the discussion and applications of deformable template matching.

2.6 Matching Cost Functions

Given a source image I and a template image T of size $M \times N$, the template-matching problem is to find the best match of T from I with minimum distortion or maximum correlation.

2.6.1 Popular Matching Cost Functions

The sum of absolute differences (SAD), the sum of squared differences (SSD), cross correlation (CC), and the normalized cross correlation (NCC) are among the most popular matching functions.

$$\text{SAD}(x,y) = \sum_{i=1}^{M} \sum_{j=1}^{N} | \text{T}(i,j) - \text{I}(x+i, y+j) |, \qquad (2.7)$$

$$\text{SSD}(x,y) = \sum_{i=1}^{M} \sum_{j=1}^{N} [\text{T}(i,j) - \text{I}(x+i, y+j)]^2, \qquad (2.8)$$

$$\text{CC}(x,y) = \sum_{i=1}^{M} \sum_{j=1}^{N} \text{T}(i,j)\text{I}(x+i, y+j), \qquad (2.9)$$

(a)
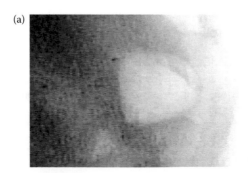

(b)
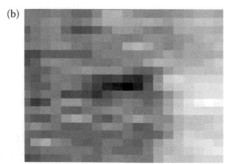

(c)
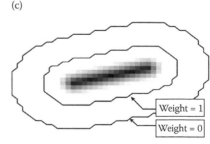

FIGURE 2.3   The object template for (a) objects within an image; (b) obtained from the input image directly (empirical); (c) constructed from a predetermined model (model-based).

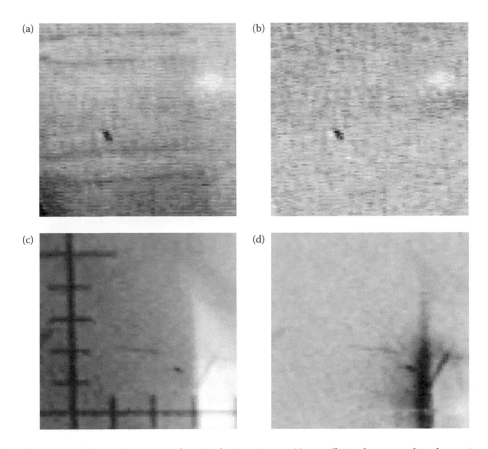

FIGURE 2.4 Motion enhancement. The stationary vertebrae in the input image (a) are effectively removed in the motion enhanced image (b). However, the same enhancement approach does not work quite so effectively as shown in (c) because the background stomach gas moves together with the marker (d).

$$\text{NCC}(x,y) = \frac{\sum_{i=1}^{M}\sum_{j=1}^{N} T(i,j) I(x+i, y+j)}{\sqrt{\sum_{i=1}^{M}\sum_{j=1}^{N} I(x+i, y+j)^2}\sqrt{\sum_{i=1}^{M}\sum_{j=1}^{N} T(i,j)^2}}.$$

(2.10)

In contrast to SAD, SSD, and CC, NCC accounts for gain differences (a multiplicative change) in the matching window because of normalization in the denominator.

The zero-mean versions of the above matching cost functions are used to compensate for a constant offset (bias) in the change of pixel values:

$$\text{ZSAD}(x,y) = \sum_{i=1}^{M}\sum_{j=1}^{N} |[T(i,j) - \widehat{T}] - [I(x+i, y+j) - \widehat{I}]|,$$

(2.11)

$$\text{ZSSD}(x,y) = \sum_{i=1}^{M}\sum_{j=1}^{N} [(T(i,j) - \widehat{T}) - (I(x+i, y+j) - \widehat{I})]^2,$$

(2.12)

$$\text{ZCC}(x,y) = \sum_{i=1}^{M}\sum_{j=1}^{N} [T(i,j) - \widehat{T}] \, [I(x+i, y+j) - \widehat{I}],$$

(2.13)

$$\text{ZNCC}(x,y) = \frac{\sum_{i=1}^{M}\sum_{j=1}^{N} [T(i,j) - \widehat{T}] \, [I(x+i, y+j) - \widehat{I}]}{\sqrt{\sum_{i=1}^{M}\sum_{j=1}^{N} [I(x+i, y+j) - \widehat{I}]^2}\sqrt{\sum_{i=1}^{M}\sum_{j=1}^{N} [T(i,j) - \widehat{T}]^2}}.$$

(2.14)

ZNCC is the only matching cost function that can compensate for both gain and offset within the matching window. In terms of computation cost, NCC and ZNCC have the highest cost. However, none of the above mentioned matching cost functions are robust, and a single outlying pixel can distort them arbitrarily.

2.6.2 Robust Matching Cost Functions

One simple robust matching cost function is the maximum of absolute difference (MAD) and its corresponding zero-mean version:

$$\text{MAD}(x,y) = \max_{\substack{i=1\ldots M, \\ j=1\ldots N}} (|T(i,j) - I(x+i, y+j)|)$$

(2.15)

$$\text{ZMAD}(x,y) = \max_{\substack{i=1\ldots M, \\ j=1\ldots N}} (|(T(i,j) - \widehat{T}) - (I(x+i, y+j) - \widehat{I})|).$$

(2.16)

MAD and ZMAD are less sensitive to outliers and thus more robust. They are also easier to compute than NCC and ZNCC.

Robust M-estimators are also used, and they are more resilient to outliers. The characteristic of robust estimators is that they cause outliers to contribute less weight compared to the

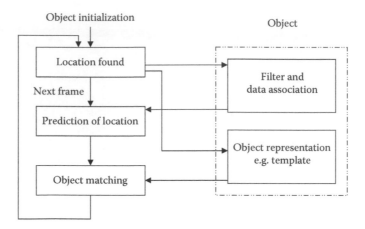

FIGURE 2.5 The key steps in an image-based tracking algorithm.

inliers. Most M-estimators include a parameter that needs to be set beforehand: the point at which measurements must be considered as outliers. The ideal value for this threshold can vary depending on image contrast and noise level, and setting it in a correlation-based framework is often done empirically. It is desirable to have a universal measure of correlation independent of absolute intensity scale and experimental conditions.

2.7 Tracking and Prediction

The key steps that are common in almost all current image-based tracking systems are shown in Figure 2.5. The tracking algorithm begins when the initial object locations are identified either through user interaction or by an object detection module. The object representations found at the locations of the objects are stored. As mentioned previously, these can be predefined templates or other image features. In the next frame, the filtering and association module makes a prediction of the possible locations for each object being tracked. Then in the object-matching step, the stored templates are used to search for the object within a search area around the predicted locations. The best-matched locations (based on the matching cost) are determined, and one or more of them are chosen to be the best location estimate. The best matching locations are used as feedback to correct the filtering and association module, and the algorithm continues on to the next frame repeating the same process.

In the previous sections, we have discussed object representation and matching using templates. Now we look at some standard filtering, prediction, and data association algorithms. We use X to represent the location of the object, which is called the state of the object in the context of most tracking algorithms. The change of state over time is given by the following equation:

$$X_{t+1} = \phi(X_t) + W_t, \tag{2.17}$$

where X_t is the state vector at time t, ϕ is the function that governs the dynamics of the moving object and W_t is associated

noise process. The image observation of X_t can be written as

$$Z_t = \mathcal{C}(X_t) + V_t, \tag{2.18}$$

where Z_t is the current image observation, C is the image measurement function based on the current state X_t, and V_t is the associated noise process.

Depending on how ϕ and C are modeled, there are many tracking algorithms in the computer vision literature. We broadly categorize some standard tracking algorithms based on whether they are a single object tracking or multiobject tracking system.

2.7.1 Single Object Tracking

For the case of tracking a single object, if both ϕ and C are linear functions and the object state X and the noise follows a Gaussian distribution, then the optimal estimate of X_t is given by the Kalman Filter [Kal60]. However, in many cases, the object state X is not assumed to be Gaussian. In these situations, X over time can be estimated using particle filters [GSS93].

2.7.1.1 Kalman Filters

A Kalman filter is used to estimate the state of linear system where the noise is assumed to be distributed as a Gaussian. There are two steps in the standard Kalman filter: prediction and correction based on image observation.

The prediction step uses the modeled dynamic process to predict the new state X_{t+1} from the previous state X_t. The object dynamics ϕ in Equation 2.17 is replaced with a state transition matrix D:

$$\overline{X}_{t+1} = DX_t + W \quad \text{and} \tag{2.19}$$

$$\overline{\Sigma}_{t+1} = D\Sigma_t D^T + Q_t, \tag{2.20}$$

where $\overline{X}_{t+1}$ and $\overline{\Sigma}_{t+1}$ are the state and the covariance of the predictions at time $t+1$. Q_t is covariance matrix of the noise process W.

The correction step updates the state estimate $\bar{X}_{t+1}$ and $\bar{\Sigma}_{t+1}$ from the image observation Z_{t+1}:

$$K_{t+1} = \bar{\Sigma}_{t+1} M^T [M \bar{\Sigma}_{t+1} M^T + R_{t+1}]^{-1}, \qquad (2.21)$$

$$X_{t+1} = \bar{X}_{t+1} + K_{t+1} \underbrace{[Z_{t+1} - M \bar{X}_{t+1}]}_{v}, \qquad (2.22)$$

$$\Sigma_{t+1} = \bar{\Sigma}_{t+1} - K_{t+1} M \bar{\Sigma}_{t+1}. \qquad (2.23)$$

Here, M is the matrix form of the measurement function C, K is the Kalman gain, and R_{t+1} is the covariance associated with the image observation (measurement) process V. v is called the innovation. If template matching is used to approximate the correction step, then the updated state estimate is given by the best match obtained using a local search window around the predicted location $\bar{X}_t$.

A major limitation of the Kalman filter is the assumption that the uncertainties in the state variables are distributed as a Gaussian. As a result, the Kalman filter will perform poorly if the actual distribution of the state variables is multimodal. Particle filters overcome this limitation by using a sampling-based approach.

2.7.1.2 Particle Filters

In particle filtering, the conditional state density $P(X_t|Z_t)$ is represented by a set of N samples (particles) $\{s_t^1, \ldots, s_t^N\}$ with corresponding weights $\{\pi_t^1, \ldots, \pi_t^N\}$. The weights indicate the importance of each particle. At time t, new particles $P_t = \{(s_t^1, w_t^1), \ldots, (s_t^N, w_t^N)\}$ are sampled from the existing particles $P_{t-1} = \{(s_{t-1}^1, w_{t-1}^1), \ldots, (s_{t-1}^N, w_{t-1}^N)\}$ at $t-1$ based on a sampling scheme. Various sampling schemes have been investigated in the literature [MB99, DBR00, ST01, SR01, SBIM01].

Using the new particles P_t, the new object state (e.g. position) can be obtained by doing a weighted combination of all the particles. That is, if we set $X_0 = s_0^n$, then $X_t = \sum_{n=1}^{N} \pi_t^n s_t^n$. The particle filter-based tracking system can be initialized either manually by setting $P_0 = \{s_0^1, 1/N), \ldots, (s_0^N, 1/N)\}$ or through some detection algorithm. One problem with most of the particle filter-based tracking algorithms is sample degeneration. This happens when a large number of the samples have very low weights that are close to 0 and a few samples have weights close to 1. A remedy to this problem is to reject samples that carry very low weights and resample in order to have sufficient number of particles to approximate the state density $P(X_t|Z_t)$. The advantage of particle filter-based tracking systems is that the state density distribution need not be Gaussian. Furthermore, good results have been obtained in varied problem domains [MB99, ST01, SR01] as long as there is a sufficient number of good particles to approximate the state distribution.

2.7.2 Multiobject Tracking

Both Kalman filters and particle filters are single object tracking systems. Tracking multiple objects requires a joint solution of data association and state estimation. Specifically, when a Kalman filter or particle filter is used to track multiple objects in the image sequence, one needs to decide how to label the image observations so that they are in correspondence with the correct object. This is called data association. One simple way to achieve this is to use the nearest neighbor approach. If we have labeled image observations at t–1, then the image observation at time t which is closest to the previously labeled observation will have the same label. However, if the objects are too close to each other and have similar appearance, then it is very likely that the labeling is wrong. An incorrectly labeled observation will often cause the filter to diverge. An excellent review of common data association techniques can be found in the book by Bar-Shalom and Foreman [BS88]. Joint probability data association filtering (JPDAF) [FBSS83] and multiple hypothesis tracking (MHT) [Rei79] are two widely used approaches to tackle the data association problem in multiple object tracking. We briefly describe these two methods in the following section.

2.7.2.1 Joint Probability Data Association Filter

Let the track of an object in the scene be a sequence of image observations (e.g. blobs or blocks in the image) that are from the same object. Suppose we have N tracks and, at time t, we have K image observations, $Z_t = \{z_t^1, \ldots, z_t^K\}$. Let L be the set of possible labels. Assume that the number of objects being tracked does not change over time. Let $v_{k,n}$ be the innovation (from Equation 2.22 in the discussion of Kalman filter) associated with the track n due to the image observation z_t^k. The JPDAF associates all measurements with each track, and the combined weighted innovation is given by

$$v^n = \sum_{k=1}^{K} \beta_k^n v_{k,n}, \qquad (2.24)$$

where β_k^n is the posterior probability that the image observation z_t^k originated from the track labeled n. The weighted innovation given in Equation 2.24 can be plugged into the Kalman filter update equations (Equation 2.22) for each track n.

JPDAF is unable to handle instantiation of new tracks or termination of old tracks because of objects entering and leaving the scene. The MHT algorithm is able to handle a variable number of tracks and thus can overcome this limitation in JPDAF.

2.7.2.2 Multiple Hypothesis Tracking

In differential tracking, labels are assigned based on the state of the previous image; there is a finite chance that the assignment of labels to objects using only two image frames is incorrect. Better tracking results can be obtained if the label assignment is deferred until several frames have been examined. The MHT approach follows several label assignment hypotheses for each object at each time instance [Rei79]. The final track of the object is the most likely set of labels, given the time period of observation. MHT is able to handle the creation and termination of tracks caused by objects that enter or leave the scene. Furthermore, it can handle missing data caused by occlusion.

MHT is an iterative algorithm. Iteration begins with a set of current track hypotheses. Each hypothesis is a collection of disjoint tracks. For every hypothesis, a prediction of the object's position in the next frame is made. The predictions are then matched with actual image observations. A set of associations are established for each hypothesis based on the matching cost, which introduces new hypotheses for the next iteration. Each new hypothesis represents a new set of tracks based on the current image observations.

MHT makes associations in a deterministic manner and exhaustively enumerates all possible label assignments. To reduce the exponential cost in both computation time and memory, various heuristics have been proposed [Kur83, CH96, ORS04] to select and maintain only a subset of all possible track assignments.

References

[BS88] Y. Bar-Shalom. *Tracking and Data Association.* Academic Press Inc., Waltham, MA, 1988.

[CFJ08] V. Cheung, B.J. Frey, and N. Jojic. Video epitomes. *International Journal of Computer Vision (IJCV)*, 76(2):141–152, 2008.

[CH96] I.J. Cox and S.L. Hingorani. An efficient implementation of Reid's multiple hypothesis tracking algorithm and its evaluation for the purpose of visual tracking. *IEEE Trans. on Pattern Analysis and Machine Intelligence (PAMI)*, 18(2):138–150, 1996.

[DBR00] J. Deutscher, A. Blake, and I. Reid. Articulated body motion capture by annealed particle filtering. In *Proc. IEEE Conf. on Computer Vision and Pattern Recognition (CVPR)*, pp. 126–133, 2000.

[FBSS83] T. Fortmann, Y. Bar-Shalom, and M. Scheffe. Sonar tracking of multiple targets using joint probabilistic data association. *IEEE Journal of Ocean Engineering*, 8(3):173–184, 1983.

[GSS93] N.J. Gordon, D.J. Salmond, and A.F.M. Smith. Novel approach to nonlinear/non-Gaussian Bayesian state estimation. In *IEE Proceedings F on Radar and Signal Processing*, 140:107–113, 1993.

[JZL96] A.K. Jain, Y. Zhong, and S. Lakshmanan. Object matching using deformable templates. *IEEE Trans. on Pattern Analysis and Machine Intelligence (PAMI)*, 18(3):267–278, 1996.

[Kal60] R.E. Kalman. A new approach to linear filtering and prediction problems. *Transaction of ASME—Journal of Basic Engineering*, pp. 35–45, 1960.

[Kur83] T. Kurien. Issues in the design of practical multitarget tracking algorithms. In *Multitarget-Multisensor Tracking: Advanced Applications*, Y. Bar-Shalom, ed., pp. 43–83, Artech House, Norwood, MA, 1983.

[MB99] J. MacCormick and A. Blake. A probabilistic exclusion principle for tracking multiple objects. In *Proc. IEEE International Conf. on Computer Vision (ICCV)*, pp. 572–578, 1999.

[ORS04] S. Oh, S. Russell, and S. Sastry. Markov chain Monte Carlo data association for general multiple-target tracking problems. In *Proc. IEEE Conference on Decision and Control*, pp. 735–742, 2004.

[Rei79] D.B. Reid. An algorithm for tracking multiple targets. *IEEE Trans. Autom. Control*, 24(6):543–854, 1979.

[SBIM01] J. Sullivan, A. Blake, M. Isard, and J. MacCormick. Bayesian object localization in images. *International Journal of Computer Vision (IJCV)*, 44(2):111–135, 2001.

[SCI05] E. Shechtman, Y. Caspi, and M. Irani. Space-time super-resolution. *IEEE Trans. on Pattern Analysis and Machine Intelligence (PAMI)*, 27(4):531–545, 2005.

[SL01] S. Sclaroff and L.-F. Liu. Deformable shape detection and description via model-based region grouping. *IEEE Trans. on Pattern Analysis and Machine Intelligence (PAMI)*, 23(5):475–489, 2001.

[SR01] J. Sullivan and J. Rittscher. Guiding random particles by deterministic search. In *Proc. IEEE International Conf. on Computer Vision (ICCV)*, pp. 323–330, 2001.

[ST01] C. Sminchisescu and B. Triggs. Covariance scaled sampling for monocular 3D body tracking. In *Proc. IEEE Conf. on Computer Vision and Pattern Recognition (CVPR)*, pp. 447–454, 2001.

[ZJDJ00] Y. Zhong, A.K. Jain, and M.-P. Dubuisson-Jolly. Object tracking using deformable templates. *IEEE Trans. on Pattern Analysis and Machine Intelligence (PAMI)*, 22(5):544–549, 2000.

3

Respiratory Gating

3.1	Overview	21
3.2	Historical Development	21
3.3	Planning and Delivering a Gated Treatment	22
3.4	Candidate Patients and Treatment Sites	22
3.5	Candidate Breathing Behaviors and Respiratory Maneuvers	22

Assessing Regularity and Stability of Respiration • Breath Hold
• Feedback-Guided Free Breathing • Motion Restriction

3.6	Simulating and Planning a Gated Treatment	24
3.7	Delivering a Gated Treatment	25
3.8	Gating Quality Assurance and Control	26

Commissioning a Gating System • Routine Quality Assurance

3.9	Limitations and Future Developments	27
	References	27

Geoffrey D. Hugo

Martin J. Murphy

3.1 Overview

Intrafraction motion adaptation in radiation therapy divides broadly into two categories: temporal and spatial adaptation. Spatial adaptation (tracking) involves changing the position of the beam or the patient in response to motion. Temporal adaptation (gating) involves turning a stationary beam on and off to compensate target motion. Simply put, the treatment beam defines a line of sight. The radiation is only turned on (gated) when the target is in the line of sight.

Gating is practical (and useful) only for regular periodic motion such as respiration. Although simple in concept, a complete gated treatment involves five sophisticated tasks: (1) assessing the target motion before treatment, (2) planning for it, (3) detecting it during treatment, (4) coordinating the planned beam gate with the observed motion, and (5) verifying that the treatment is proceeding as planned. There is, in addition, the possibility of influencing the patient's breathing to further regularize it.

The tasks of target motion detection during treatment and its quality assurance are common to all motion-adaptive techniques. Therefore they are discussed in detail in separate chapters. The principal focus of this chapter will be the process of planning a gated treatment and then managing its delivery. We will focus on patient selection, pretreatment motion assessment, respiration management, and treatment planning. Our main concern in the delivery process will be the coordination of the intrafraction target tracking data with the target motion observed and planned for in the treatment planning process.

3.2 Historical Development

The practice of gating a radiation beam synchronously with breathing was first introduced to improve diagnostic x-ray imaging (Jones 1982; Frohlich and Dohring 1985). In particular, the extended period of time required for computed tomography demanded a way to mitigate motion artifacts without breath-holding.

The merit of gating a therapeutic radiation beam in response to tumor motion was under active discussion by the mid 1980s (Rekonen and Toivonen 1985; Willett et al 1987; Suit et al 1988). Actual implementation in the clinic happened first in Japan (Ohara et al 1989; Inada et al 1992; Okumura et al 1994; Minohara et al 2000). However, by the mid-1990s the practice was being developed at a number of U.S. and European institutions (see, e.g., Kubo and Hill 1996). The separability of the target detection and beam gating technologies resulted in a wide variety of system configurations using, for example, x-ray imaging, optical tracking, spirometry instruments, and strain gauges for motion detection connected to a variety of beam delivery systems, including proton accelerators (Inada et al 1992; Okumura et al 1994), gantry-mounted x-ray linacs (Ohara et al 1989), heavy ion beams (Minohara et al 2000), and robotically maneuvered x-ray linacs (Murphy and Cox 1996).

These initial gating systems relied on the same motion detection technology both for delivering a treatment and gating a diagnostic x-ray study for generating an image for treatment planning. The transition of "4D" imaging techniques from diagnostic radiology to radiation oncology in the early 2000s

(Ford et al 2003; Low et al 2003; Vedam et al 2003) allowed the mechanism for acquisition of planning images to be decoupled from the treatment delivery technology. The development of virtual motion management planning techniques followed, which has allowed customization and refinement of the treatment plan for a particular patient.

3.3 Planning and Delivering a Gated Treatment

Developing and delivering a gated treatment plan proceeds in several steps. First, one decides if the treatment site and patient traits are amenable to gating. Here a key consideration is whether the target motion is observable, significant, and regular. Secondly, if gating is indicated, then one must consider if it is practical and feasible to further regularize the patient's breathing in some way. Thirdly, one must define the gating parameters based on the motion observed in the planning images. Lastly, one must coordinate the target motion information acquired during the planning process with the real-time motion data acquired during treatment and generate a properly synchronized gate.

3.4 Candidate Patients and Treatment Sites

Respiration generally influences treatment sites in the thorax and upper abdomen. Common respiratory-influenced sites in radiation therapy include the breast, liver, lung, pancreas, and esophagus. Although targets located in these sites are often prone to respiration-induced motion, no strong predictors for the magnitude of motion have been found. It is not possible to decide only on the basis of target location if motion management is indicated. Instead, the magnitude and shape of motion should be measured by imaging each patient individually with treatment decisions based partially on this assessment.

The imaging technique selected for measuring target motion should be capable of resolving either the target directly or a strong surrogate. For targets embedded in the liver, the liver itself or the hemidiaphragm is often a good surrogate. Similarly, for breast tumors, the skin surface or chest wall may be a representative of target motion. For the majority of respiratory sites, 4D computed tomography (4DCT) is the most commonly used method, as it possesses good spatial and temporal resolution for this task and is often capable of resolving the target directly. Average target excursion of less than 5 mm from the end of exhalation to end of inhalation has been shown to have a negligible effect on the dose distribution (van Herk 2004). In this case, the additional technical requirements of gating likely have limited the benefits for the patient, and we recommend excluding such patients from gated radiotherapy.

In addition to the average motion of the target, one needs to consider other factors in formulating an appropriate motion management strategy. Targets adjacent to risk structures may benefit from the higher precision afforded by gating, which potentially reduces the risk of toxicity to the normal organ. The treatment goal (i.e., curative or palliative) should be considered, as curative schedules—with generally higher prescription dose—require more aggressive interventions for reducing normal tissue dose. Patient tolerance to the potentially longer treatment times and more participatory nature of gating (particularly as described below for breath-hold and feedback-guided methods) should be factored into the selection process. To develop a more consistent practice, we recommend that institutions develop a set of written criteria for patient selection based on these factors.

3.5 Candidate Breathing Behaviors and Respiratory Maneuvers

3.5.1 Assessing Regularity and Stability of Respiration

As the beam and patient are not adjusted during delivery, gating relies on the regularity and reproducibility of the respiratory cycle, and hence the target motion, for accurate and efficient delivery. The basic premise of gated radiotherapy is for the radiation beam to be delivered only when the target is at a predefined position or phase in the respiratory cycle. If the target does not return to the same position with each cycle, this assumption is violated and treatment delivery may be affected. Figure 3.1 demonstrates an example of how such irregular motion, depending on the type of gating, may reduce either the efficiency or accuracy of delivery.

Irregularity of the respiratory cycle can be divided into two major types: (1) change in the respiratory pattern during a fraction, and (2) change in the respiratory pattern between fractions. In the remainder of this section, we discuss methods to manage irregular respiration during a single fraction. Change in the respiratory pattern between fractions can be managed concurrently with other fraction-to-fraction anatomical change. Management methods for these types of change are therefore deferred to Section 3.7 on gated delivery.

Determination of the regularity of a patient's respiration should be part of the selection process and can be made during the same imaging session that was used to measure the motion magnitude. Common approaches include evaluating the periodicity and cycle-to-cycle reproducibility of a continually acquired surrogate signal (e.g., an optical system measuring abdominal height or a strain gauge). 4DCT images can also be inspected for artifacts as an indirect metric of regularity (see Figure 3.2), as these artifacts are often due to irregular respiration; however, caution should be exercised with this method, as other causes for artifacts exist (e.g., incorrect table pitch for a helical 4DCT acquisition). Mild irregularity can be managed in treatment planning (see Section 3.6 below on gated planning). Larger irregularity such as cycle-to-cycle variation that substantially occludes the underlying periodic breathing pattern, drifting in the average target position over

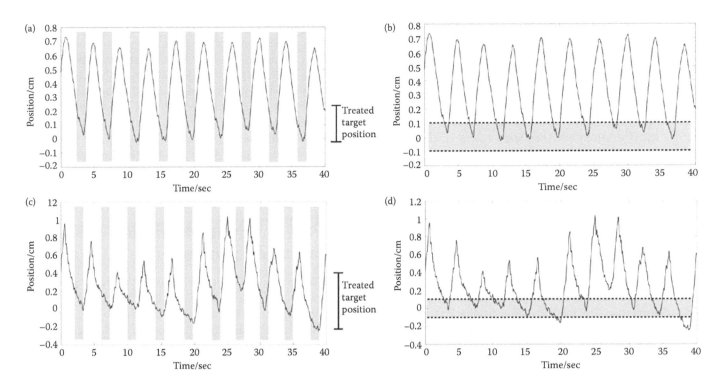

FIGURE 3.1 Breathing traces for different types of gating during regular and irregular respiration. (a) Phase gating, regular respiration. (b) Amplitude gating, regular respiration. (c) Phase gating, irregular respiration. (d) Amplitude gating, irregular respiration. The gray areas correspond with beam on during the gate. Note that with phase gating, irregular respiration produces larger residual error than with regular respiration. For amplitude gating, the residual error for regular and irregular respiration is similar, but the duty cycle is shorter for irregular respiration, prolonging treatment.

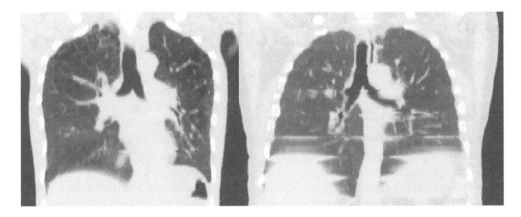

FIGURE 3.2 Coronal sections from single 4DCT phase images demonstrating minimal and heavy artifact related to respiratory regularity. Left: Patient with minimal artifact due to regular respiration. Right: Patient with heavy artifact due to irregular respiration.

time, persistent coughing, or continual change between irregular and regular breathing should exclude a patient from free-breathing gating.

3.5.2 Breath Hold

Although gating generally implies delivery during free breathing, a more general definition includes any motion-adaptive method that relies on a static beam-to-patient relationship.

breath-hold methods fall under the general definition of gating if one thinks of the breath-hold condition itself as a gate.

To further this discussion, we first introduce the concept of duty cycle. Duty cycle is the fraction of overall treatment time during which the beam is on. For perfectly regular, periodic breathing, the duty cycle is simply the time per cycle during which the beam is on divided by the respiration period. Realistically, the duty cycle can be calculated by measuring the beam on time and dividing it by the total time to deliver the beam.

Breath-hold methods may be thought of as an artificial means to extend the duty cycle by maximizing the length of the gate while keeping target motion small within the gate. The major development in efficient breath-hold methods took place in the late 1990s (Hanley et al 1999; Wong et al 1999; Murphy et al 2002). Deep-inspiration and moderate deep-inspiration breath-hold methods were based on a modified vital capacity maneuver in which the patient exhales entirely and then takes a deep inhale and holds. The process is voluntary, so concerns about reproducibility have resulted in the development of feedback systems to aid the patient in achieving a consistent breath-hold level (Nelson et al 2005). Another approach is a computer-controlled, rather than voluntary, breath hold for improving reproducibility. This approach, termed active breathing control, uses a spirometer and an attached valve to measure and induce breath hold at a predefined lung volume. Currently, both deep and normal inspiration breath-hold techniques are common. End-of-exhalation breath hold is uncommon as the technique is more difficult for the patient to tolerate. Also, end-of-inspiration (particularly deep inspiration) breath hold may have some potential benefit in reducing normal tissue dose compared to end-of-exhalation breath hold, as the larger lung volume reduces mean lung dose and may move targets away from risk structures (Partridge et al 2009).

Considerations for breath-hold radiotherapy are similar to free breathing gating in that an extended treatment time is required and the technique should only be tried if there is an expected benefit to reduced target motion. Furthermore, patients must be capable of repetitive breath holds. For example, a 2-Gy fraction on a modern gantry-mounted linear accelerator requires 3–10 breath holds of 10–30 seconds duration, depending on pretreatment imaging, number of beams, and dose rate. There are some recent evidences that repeated end-of-normal-inspiration breath hold of short (5–10 Seconds) duration is more tolerable (Glide-Hurst et al 2010).

3.5.3 Feedback-Guided Free Breathing

For patients with irregular breathing, who are not candidates for breath hold, another option is to attempt to regularize the respiration pattern. Audio coaching consists of a predefined, periodic audio signal that the patient follows to control the period of respiration (Kini et al 2003). Audio–visual feedback-based systems have also been developed (Kini et al 2003; George et al 2006; Venkat et al 2008), combining the audio coaching with a visual system. Here, the patient views a trace of their breathing pattern and attempts to keep the trace within a predefined trajectory. The predefined pattern can be patient specific and set during an initial training session. To date, there is little data on the clinical performance of feedback and coaching systems to regularize target motion.

3.5.4 Motion Restriction

Besides the techniques that involve forcing a breath hold or attempting to guide free breathing, another technique is to attempt to dampen target motion by constraining the anatomy. The most common approach uses a screw-driven plate to compress the upper abdomen. Such systems have been shown to reduce target motion by 50%, but only with relatively high compression forces (Heinzerling et al 2008). One issue to consider with such restriction systems is that the patient must still breathe a similar volume of air as during free breathing. If the abdomen—and thus the diaphragm—is restricted, the patient then recruits intercostal muscles to breathe more prominently using the chest wall. As such, abdominal compression may only reduce motion for targets near the diaphragm, such as liver and lower lobe lung tumors.

3.6 Simulating and Planning a Gated Treatment

Once the patient has been selected for an appropriate gating strategy (e.g., free breathing or breath hold) and necessary breathing regularization has been applied, the baseline position and motion of the target, risk structures, and other anatomy must be assessed during simulation. All images for planning should be acquired under the same regularization conditions as will be used during treatment. For free-breathing gating, a 4DCT can be used to virtually plan the gated delivery. If 4DCT is unavailable, one can acquire a prospectively gated CT, where each slice acquisition begins after receiving a trigger when the patient enters the gate. Prospective gating provides a single, static CT scan. For breath hold, several breath-hold CTs should be acquired to enable estimation of intrafractional reproducibility of the breath hold.

The ideal target detection system for gating would gate the beam based on real-time, continuous, three-dimensional observation of the tumor itself. Fast, robust imaging systems that can perform such a task with acceptable patient dose are not available. Instead, surrogates of the target position are generally employed. Fiducial markers, either active or passive (Shirato et al 2000), implanted in or near the target have been used. However, there are still some locations where it is difficult to implant or locations that carry more than minimal risk of side effects from the implantation process.

The workhorse of respiration surrogates is the external respiration signal. A variety of options exist for measuring respiration from the patient surface. Abdominal strain gauges measure strain due to abdominal distension (Kubo and Hill 1996). Optical systems using an infrared light source and camera or laser range finder measure one-dimensional or three-dimensional displacement of the abdomen and/or thorax (Kubo and Hill 1996). Systems based on measuring expired and inspired air volume using spirometry or temperature changes have also been employed.

Regardless of the type of external surrogate used, the surrogate signal should be measured concurrently with simulation image acquisition. For example, if using an optical marker-based system and 4DCT, the optical signal should be acquired while 4DCT acquisition takes place, and should either be used to perform the sorting process directly for reconstruction of 4D images, or the mean signal during each 4D time bin should be measured. The correlation between surrogate signal and image is necessary to map the gating thresholds between the surrogate signal and simulation images.

Gating requires selection of the "width" and location of the gate. Several factors should be considered when choosing the gate. Increasing the gate width will increase the duty cycle, but it will also result in larger residual motion within the gate, as a larger proportion of the total target motion will be captured. A too-small gate width may result in prolonged treatment times that might increase the possibility of residual error due to loss of correlation between the surrogate signal and target position. A reproducible and stable state of respiration (and thus target position) within the gate should be selected as the gating location. States (such as end of expiration) that have relatively long periods of little motion are preferred. It is common to use a fixed width and location for all patients (e.g., 30% of the average target motion centered about the end expiration target position).

Another option is to design the gate width and location based on the 4DCT images, and map the gate from the images to the surrogate signal (Wink et al 2008). Essentially, this is the reverse of the above-described process and proceeds as follows. One identifies a subset of the 4DCT phase images about which one wishes to gate based on the desired level of residual target motion within the gating window. For example, one can use either the borders or centroid from the delineated target on each 4DCT phase image to measure expected residual target motion as a function of number of phase images captured in the gate. Next, one selects all phases that give the desired level of residual motion. The surrogate signal corresponding to each phase image is identified, and the maximum and minimum surrogate signals out of this subset are used as thresholds to define the gate width and location.

For a prospectively gated CT simulation, this image-designed gate is not possible. For prospectively gated CT, the gate location must be specified at the time of CT as it is used to trigger the CT acquisition. Two methods can be used to set the gate for this case. The simplest is to use the same surrogate signal to acquire the prospectively gated CT as will be used for gated treatment delivery. In this case, one must define the gate width and location for prospectively gated CT and employ the same gate for delivery. If the same surrogate cannot be used, one must measure the surrogate signal during CT acquisition and determine the gating width and location in the surrogate signal based on the timing of CT acquisition.

In addition to the selection of gating width and location, one must choose between amplitude and phase gating. In amplitude gating, both the gate width and location are determined based on the surrogate signal amplitude (see Figure 3.1). If the surrogate signal is considered a periodic signal, the phase is in the temporal position relative to a reference state. For example, consider one full breathing cycle. The current end of inspiration state is the 0% phase, the upcoming inspiration state is the 100% phase, and all time points in between them are normalized to

this percentage scale. One can design a gate width and location in this phase space.

Amplitude gating and phase gating result in similar, although not identical, duty cycle and residual target motion (residual motion within the gate) for perfectly periodic, regular respiration. Assuming good correlation between the surrogate and target position, amplitude gating under irregular respiration will result in less residual error than phase gating but may increase the duty cycle (see Figure 3.1). However, in this case, amplitude gating may force beam delivery at various respiration states with the same amplitude, which may result in different patient configurations from cycle to cycle. Phase gating during irregular respiration will result in delivery at the same state, but not necessarily at the same surrogate signal amplitude.

In most clinical conditions, the target will not be stationary within the gate during delivery due to a variety of factors including gate width, regularity of respiration, strength of surrogate signal/target position correlation, and imaging and delivery latencies. It is important to measure and compensate for such sources of variability by incorporating them into a safety margin. Estimates of the residual target motion due to surrogate quality and gate width are available in the literature (Gierga et al 2005). Residual target motion within the gating window can be estimated from the planning 4DCT scan or another motion-inclusive imaging method such as fluoroscopy. The distribution of residual targeting error generated from these sources can coupled with proper margin formulation (van Herk 2004) to incorporate these estimates of gating-specific uncertainty into the total safety margin.

Another widely used—but technically incorrect—method involves managing finite gate width by forming a gated internal target volume (ITV_g) using a 4DCT scan. The ITV_g is formed by taking the union of the gross tumor volume contoured on 4DCT phase images within the gate. One can then add additional margin to compensate for surrogate quality and latency (time lag between position measurement and gate activation). The ITV_g method only compensates for finite gate width and not other sources of residual error such as surrogate/target correction and latency. Furthermore, use of an ITV_g causes difficulty in generating the total safety margin for other error sources, such as setup error and delineation uncertainty. Generally, we recommend the first method described above be used to fold gating-specific uncertainties into the total margin.

3.7 Delivering a Gated Treatment

Although target trajectories are relatively stable, daily variations in the average position of lung and liver tumors in relation to the bony anatomy are large, often much larger than the respiration-induced target motion itself. Surrogate signals do not predict these changes in baseline tumor position and therefore cannot be used to update the target localization at each fraction. In fact, the surrogate/target position relationship often does not hold well between treatment days. For this reason, it is recommended that at the beginning of each treatment fraction, one should (1) update the surrogate/target position relationship, and (2) localize the target position using image guidance.

Although much evidence demonstrates this necessity, unfortunately, only a few commercial options for updating the surrogate/target position correlation at the treatment unit are available. Several vendors have systems based on fixed stereoscopic x-ray systems, which detect internal fiducial markers and update the correlation with an optical external surrogate system. 4D cone beam CT is a viable option, but it is currently limited mainly to research institutions.

If these options are not available, digital fluoroscopy or radiography using either electronic portal imaging or imaging with a separate imaging beam line should be employed to aid in both localization and to update gate width and location based on target position. This process can be performed simply by acquiring radiographs during the gate (manually, if necessary) and using these to aid in localization (Balter et al 2002); it is widely available on many modern treatment units. If using single gated radiographs, we suggest acquiring several of them during different cycles to estimate and verify residual motion and the corresponding safety margin.

During delivery, the surrogate signal should be monitored for changes in breathing pattern, which may suggest degradation of the surrogate/target position relationship. For example, coughing, signal drift, and deep breaths should terminate delivery. The surrogate/target position relationship should be measured during delivery periodically or after any of these episodes. Some vendors have systems that perform such updating automatically, but this is not a common implementation.

3.8 Gating Quality Assurance and Control

Because of the patient-specific nature of the surrogate/target position relationship, quality monitoring in gating is heavily skewed towards patient-driven quality control, as described in the previous section. However, a quality assurance program is necessary to ensure proper operation of any surrogate systems, to ensure gated beam delivery, and to ensure proper physical connection (calibration) and/or time stamping between the surrogate system and imaging systems necessary to establish the surrogate/target position relationship.

3.8.1 Commissioning a Gating System

The commissioning procedure for a gating system will vary depending on one's particular configuration of imaging, surrogate, and delivery systems. However, several key steps should be performed regardless of system configuration, including (1) validating surrogate/target position correlation measurement; (2) benchmarking treatment unit dosimetry under gated operation; and (3) benchmarking system latency and response.

A gating system must be capable of relating target position as measured by imaging to the surrogate signal. The simplest and most effective test for establishing this calibration accuracy is to use a moving phantom or object that is detectable by both the imaging system and surrogate system. For example, phantoms that mimic respiration that have embedded markers for imaging and can accommodate external reflective markers for surrogate measurement are available. Because these signals are inherently correlated by the rigid motion of the phantom, one measures the individual signals and then determines how closely the gating system establishes the appropriate correlation.

A moving phantom can also be used in a prospective manner to evaluate the latency in triggering the imaging system, which will propagate to the establishment of surrogate/target relationship. The gating system is set to a narrow gate on the surrogate signal, which corresponds to a known phantom position. Gating the imaging system, the measured phantom position in the position is compared to the nominal position. Knowing the motion pattern of the phantom, one can calculate the temporal latency from the positional difference and known velocity. All imaging systems, both for simulation and for treatment delivery, should be evaluated in this manner.

Gating requires the treatment delivery system to be capable of rapid modulation of the dose rate from zero to the clinical dose rate, and the reverse. As this process occurs for every gating (i.e., breathing) cycle, instability in beam on and off can quickly add up to reduce the accuracy of dose delivery. Treatment units by different vendors have various means of gating the treatment beam, including beam hold on the electron gun or on the radiofrequency power system. Literature on gated dosimetry is available on these various commercial systems, but it is important in commissioning to at least verify gated output against nongated (conventional) delivery under clinical conditions (i.e., a clinically relevant duty cycle and number of gating cycles). This test can be performed with an ionization chamber using the setup for monthly output verification, for example. The beam is gated and output is measured, and this result is compared to that from conventional, nongated operation.

The prospective moving phantom imaging test discussed above can be adapted to measure latency in treatment beam delivery and the effect on accuracy of dose delivery. A piece of film can be attached to the moving phantom and a narrow gate used to deliver the treatment beam to the phantom based on the surrogate signal. Beam location on the film relative to that for stationary delivery can be used to quantify system latency in delivery. Film, ion chamber, or other dosimeters can be attached to the moving phantom and used to establish dosimetric accuracy due to both latency and gated operation.

Finally, a system test should be performed to assess operation of all gated components together. In such a test, the moving phantom is treated using the same procedure as a patient: simulation, planning, and establishment of gate width and location should be performed; the surrogate signal should be used to trigger imaging and evaluate spatial and temporal accuracy of the imaging system on the treatment unit; gated delivery accuracy should be assessed either by imaging the phantom with the treatment beam and/or dosimetrically.

3.8.2 Routine Quality Assurance

Routine quality evaluation should include performance evaluations for imaging and surrogate equipment. Gated operation should be verified dosimetrically using the modified monthly output check discussed above. The system test should be repeated after system repairs, upgrades, or replacement and on a routine schedule.

3.9 Limitations and Future Developments

Since the introduction of gating in the 1980s, much has been learned about the characteristics of respiration and their effect on target motion. This knowledge has resulted in better surrogate systems and better target detection systems that have contributed to more accurate and precise gating systems. However, several key limitations of modern gating systems remain to be solved.

Recent imaging studies in locally advanced non-small-cell lung cancer have recognized that target structures do not necessarily move in rigid configurations. A primary tumor and involved lymph nodes often have different trajectories, motion magnitudes, and baseline positions everyday. Strategies for treating multiple targets with gating (or with most motion management techniques for that matter) are relatively undeveloped. As the majority of radiation therapy lung cancer patients present with involved lymph nodes that require simultaneous treatment with the primary tumor, this is an important unsolved problem.

Clinically deliverable dose rates are increasing in modern treatment units, particularly with the advent of flattening-filter-free accelerators. With high dose rates, shorter treatments are possible, but errors over the fraction have less of a tendency to "wash out." It may soon be possible to deliver an entire fraction even with a small number of breath holds or gates. In this situation, very high correlation between the surrogate and target position must be maintained at each gate; a single cycle with large surrogate/target correspondence error would contribute substantially to dosimetric error. Thus, continued development of high-quality surrogates, strategies for updating surrogates with image-derived target position information, and prediction of future target position several seconds into the future is warranted.

It has been well documented by many observers that the surrogate/target relationship often changes during a delivery fraction, which in turn requires intrafraction recalibration of the correlation. Nevertheless, the means to perform this recalibration using various imaging modalities are only available in a few commercial systems. For gating to realize its full potential, it is important that intrafraction target position monitoring become a standard feature.

Gated radiotherapy has progressed to be a sophisticated strategy for high-precision targeting of moving targets. High-quality surrogates for target position are available, and more efficient and precise methods for minimizing residual target

motion within the gating window are ready for clinical testing and evaluation. Although available for routine clinical use today, controlled clinical trials in large patient cohorts will help demonstrate clear evidence of the benefit of motion management technology.

References

Balter, J. M., K. K. Brock, D. W. Litzenberg, et al. 2002. Daily targeting of intra-hepatic tumors for radiotherapy. *Int. J. Radiat. Oncol. Biol. Phys.*, 52(1): 266–71.

Ford, E. C., G. S. Mageras, E. Yorke, and C. C. Ling. 2003. Respiration-correlated spiral CT: A method of measuring respiratory-induced anatomic motion for radiation treatment planning. *Med. Phys.*, 30(1): 88–97.

Frohlich, H. and W. Dohring. 1985. A simple device for breath-level monitoring during CT. *Radiology*, 156(1): 235.

George, R., T. D. Chung, S. S. Vedam, et al. 2006. Audio-visual biofeedback for respiratory-gated radiotherapy: Impact of audio instruction and audio-visual biofeedback on respiratory-gated radiotherapy. *Int. J. Radiat. Oncol. Biol. Phys.*, 65(3): 924–33.

Gierga, D. P., J. Brewer, G. C. Sharp, et al. 2005. The correlation between internal and external markers for abdominal tumors: Implications for respiratory gating. *Int. J. Radiat. Oncol. Biol. Phys.*, 61(5): 1551–58.

Glide-Hurst, C. K., E. Gopan, and G. D. Hugo. 2010. Anatomic and pathologic variability during radiotherapy for a hybrid active breath-hold gating technique. *Int. J. Radiat. Oncol. Biol. Phys.*, 77(3): 910–17.

Hanley, J., M. M. Debois, D. Mah, et al. 1999. Deep inspiration breath-hold technique for lung tumors: The potential value of target immobilization and reduced lung density in dose escalation. *Int. J. Radiat. Oncol. Biol. Phys.*, 45(3): 603–11.

Heinzerling, J. H., J. F. Anderson, L. Papiez, et al. 2008. Four-dimensional computed tomography scan analysis of tumor and organ motion at varying levels of abdominal compression during stereotactic treatment of lung and liver. *Int. J. Radiat. Oncol. Biol. Phys.*, 70(5): 1571–78.

Inada, T., H. Tsuji, Y. Hayakawa, A. Maruhashi, and H. Tsujii. 1992. Proton irradiation synchronized with respiratory cycle. *Nippon Acta Radiol.*, 52(8): 1161–67.

Jones, K. R. 1982. A respiration monitor for use with CT body scanning and other imaging techniques. *Br. J. Radiol.*, 55(655): 530–33.

Kini, V. R., S. S. Vedam, P. J. Keall, et al. 2003. Patient training in respiratory-gated radiotherapy. *Med. Dosim.*, 28(1): 7–11.

Kubo, H. D. and B. C. Hill. 1996. Respiration gated radiotherapy treatment: A technical study. *Phys. Med. Biol.*, 41(1): 83–91.

Low, D. A., M. Nystrom, E. Kalinin, et al. 2003. A method for the reconstruction of four-dimensional synchronized CT scans acquired during free breathing. *Med. Phys.*, 30(6): 1254–63.

Minohara, S., T. Kanai, M. Endo, K. Noda and M. Kanazawa. 2000. Respiratory gated irradiation system for heavy-ion radiotherapy. *Int. J. Radiat. Oncol. Biol. Phys.*, 47(4): 1097–103.

Murphy, M. J. and R. S. Cox. 1996. The accuracy of dose localization for an image-guided frameless radiosurgery system.
Med. Phys., 23(12): 2043–9.

Murphy, M. J., D. Martin, R. Whyte, et al. 2002. The effectiveness
of breath-holding to stabilize lung and pancreas tumors
during radiosurgery. *Int. J. Radiat. Oncol. Biol. Phys.*, 53(2):
475–82.

Nelson, C., G. Starkschall, P. Balter, et al. 2005. Respiration-
correlated treatment delivery using feedback-guided breath
hold: A technical study. *Med. Phys.*, 32(1): 175–81.

Ohara, K., T. Okumura, M. Akisada, et al. 1989. Irradiation synchronized with respiration gate. *Int. J. Radiat. Oncol. Biol.
Phys.*, 17(4): 853–57.

Okumura, T., H. Tsuji, and Y. Hayakawa. 1994. Respiration-
gated irradiation system for proton radiotherapy. *11th
International Conference on the Use of Computers in
Radiation Therapy.*

Partridge, M., A. Tree, J. Brock, et al. 2009. Improvement in
tumour control probability with active breathing control
and dose escalation: A modelling study. *Radiother. Oncol.*,
91(3): 325–29.

Rekonen, A. and J. Toivonen. 1985. Breathing gated radia-
tion therapy – possibilities and need. *XIV International
Conference on Medical and Biological Engineering and VII
International Conference on Medical Physics.*

Shirato, H., S. Shimizu, T. Kunieda, et al. 2000. Physical aspects of
a real-time tumor-tracking system for gated radiotherapy.
Int. J. Radiat. Oncol. Biol. Phys., 48(4): 1187–95.

Suit, H. D., J. Becht, J. Leong, et al. 1988. Potential for improvement in radiation therapy. *Int. J. Radiat. Oncol. Biol. Phys.*,
14(4): 777–86.

van Herk, M. 2004. Errors and margins in radiotherapy. *Semin.
Radiat. Oncol.*, 14(1): 52–64.

Vedam, S. S., P. J. Keall, V. R. Kini, et al. 2003. Acquiring a four-
dimensional computed tomography dataset using an external respiratory signal. *Phys. Med. Biol.*, 48(1): 45–62.

Venkat, R. B., A. Sawant, Y. Suh, R. George and P. J. Keall. 2008.
Development and preliminary evaluation of a prototype
audiovisual biofeedback device incorporating a patient-specific guiding waveform. *Phys. Med. Biol.*, 53(11): N197–208.

Willett, C. G., R. M. Linggood, M. A. Stracher, et al. 1987. The
effect of the respiratory cycle on mediastinal and lung
dimensions in Hodgkin's disease. Implications for radiotherapy gated to respiration. *Cancer*, 60(6): 1232–37.

Wink, N. M., M. Chao, J. Antony and L. Xing. 2008. Individualized
gating windows based on four-dimensional CT information for
respiration-gated radiotherapy. *Phys. Med. Biol.*, 53(1): 165–75.

Wong, J. W., M. B. Sharpe, D. A. Jaffray, et al. 1999. The use of
active breathing control (ABC) to reduce margin for breathing motion. *Int. J. Radiat. Oncol. Biol. Phys.*, 44(4): 911–19.

The CyberKnife® Image-Guided Radiosurgery System

4.1   Adaptation to Nonperiodic Movement 29
4.2   Adaptation to Periodic (Respiratory) Motion 30
References 31

Martin J. Murphy

The CyberKnife® is the first system designed specifically for motion-adaptive radiation therapy (and the first to be put into practice). It was originally conceived as a means to perform frameless cranial radiosurgery (Adler et al 1999). In conventional radiosurgery, a steel frame is attached rigidly to the patient's skull at the time of treatment planning and remains in place until treatment has been completed. The frame serves two purposes: it keeps the skull motionless during imaging and treatment, and it provides a means to transfer the three-dimensional lesion coordinates from the planning images to the treatment setting. It (or its equivalent) is necessary because the intense single-fraction doses that are characteristic of radiosurgery require submillimeter placement precision to avoid significant harm to critical structures. Without the rigid frame, the skull can move. Even with a well-fitted facemask restraint (e.g., an AquaPlast® mask), skull movement can be as much as several millimeters (Murphy 2009), which greatly exceeds the accepted radiosurgical tolerance. To maintain submillimeter precision without a frame, it is necessary to detect and to correct for patient's movement.

The strategy for motion adaptation adopted by the CyberKnife is to move the linear accelerator (linac) itself in response to target movements. This strategy drove all of the basic design considerations for the system. The linear accelerator (a lightweight 6 MV X-band linac) is mounted on an industrial robotic arm that can position and orient the treatment beam with 6 degrees of freedom. Two ceiling mounted diagnostic x-ray tubes and two floor-mounted flat-panel detectors periodically capture x-ray images of the target anatomy, an image registration process computes the target lesion coordinates, and a real-time control loop directs the robot to adapt its beam alignment to the current target position. Figure 4.1 illustrates the system's basic configuration.

Soon after the introduction of the CyberKnife in 1994, it was realized that the system could potentially treat sites anywhere in the body using its frameless image-guided motion adaptation capabilities. In particular, elimination of the frame removed the requirement that the entire treatment dose be delivered in a single fraction. This enabled hypofractionated body radiosurgery

for spinal lesions and soft-tissue targets such as lung tumors. The image guidance and motion adaptation capabilities of the CyberKnife also introduced a new level of targeting precision for conventional fractionated radiation therapy. Consequently, the system is presently used for a comprehensive range of radiation therapy applications although its primary strength remains the performance of high-precision hypofractionated radiosurgery.

Each potential treatment application presents a distinctive scenario for target detection and tracking. Target movement can involve random variation around a mean position, systematic drift, periodic (respiratory) motion, or a combination of the three. Furthermore, treatment sites can be located and tracked with reference to radiographic features, motion surrogates, or a combination of the two. By way of example, sites within the cranium are targeted with reference to the skull surface, sites along the spine can be located either via metal fiducials implanted in the vertebrae (Murphy et al 2000; Murphy et al 2001; Ryu et al 2001) or by direct radiographic localization of the nearby vertebral bodies (Ho et al 2007), and the prostate can be localized by implanted fiducials (King et al 2009, Friedland et al 2009). In each of these cases, intrafraction motion is generally a combination of nonperiodic systematic drift combined with random fluctuations. Tumors in pancreas and lungs that move with respiration can be located using a combination of radiographic imaging of implanted fiducials and respiratory motion surrogates (Murphy et al 2000, Schweikard et al 2000, Whyte et al 2003). In selected instances, lung tumors can be observed directly in the kV x-ray images (Murphy 2004, Fu et al 2007).

4.1 Adaptation to Nonperiodic Movement

In a typical treatment fraction involving nonrespiratory motion, the CyberKnife will sequentially deliver narrowly collimated (5–60 mm diameter) pencil beams of radiation from 50 to 100 different points (nodes) on a sphere centered on the patient's treatment site. At each node, the target position is measured via

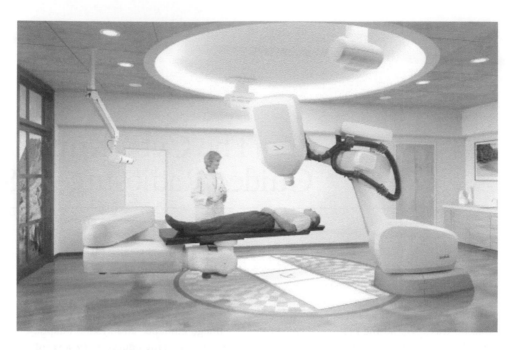

FIGURE 4.1 Basic configuration of the CyberKnife system, depicting the linear accelerator mounted to a robotic manipulator, the treatment couch, two ceiling-mounted kV x-ray tubes and their associated floor-mounted flat-panel image detectors, and the optical tracking system used to monitor continuous patient breathing movement.

x-ray imaging and the linac position is adjusted. Therefore, the nodal sphere moves with the treatment site, so the treatment planning and delivery processes are not constrained by a fixed isocenter. This provides more flexibility in treatment planning and, at the same time, introduces motion adaptation issues that are specific to the CyberKnife. Some of these issues are taken up in Chapter 8.

Random fluctuations cannot be tracked effectively, but systematic drifting can be followed. As it happens, systematic changes in target alignment are considerably more important than random fluctuations (Van Herk et al 2002). With the CyberKnife tracking system, one can make periodic adjustments of alignment during the fraction, just as if one were repeating the setup process that characterizes a sequence of daily fractions. The frequency of position measurement and correction is programmable to minimize imaging dose while maintaining the required radiosurgical precision. For cranial radiosurgery fractions lasting 20–30 minutes, it has been shown that if the patient is immobilized in a restraint such as an AquaPlast® mask, then a measurement and adaptation interval of about 1 minute is generally sufficient to maintain target alignment to within 1 mm (Murphy et al. 2003). In the absence of any intrafraction tracking, cranial position uncertainties average 3–4 mm (Murphy 2009).

4.2 Adaptation to Periodic (Respiratory) Motion

Initially, motion adaptation for lung and pancreas tumors was accomplished by breathholding (Murphy et al 2000, Whyte et al 2003). Each time the patient took and held a breath, the system took an image and adjusted the beam alignment. This conservative approach was dictated by the imaging requirements. With only a kV x-ray system available to observe the target position, tracking during free breathing would have required essentially continuous fluoroscopic imaging, as, for example, was used by the beam-gated Hokkaido system (Shirato et al 2000). However, the much longer sessions of a hypofractionated radiosurgery regimen would have required 20–30 minutes of nearly-continuous fluoroscopy, which would have delivered an unacceptable concomitant dose from imaging.

Nevertheless, it was obvious that the CyberKnife was naturally suited to tracking moving lung tumors during free respiration if a practical near-real-time source of target position data could be developed. This was accomplished by introducing an infrared optical motion tracking apparatus that operates in parallel with the x-ray imaging system. The optical system continuously monitors the movement of the patient's chest surface. This movement is correlated to the tumor's position as it is observed in a sequence of x-ray images. The chest motion is translated into tumor motion via the observed correlation function and transmitted in real time to the robot (Schweikard et al 2000). This enables the system to respond continuously to respiratory motion as it is being measured by the optical tracker. The x-ray imaging system is used periodically to update the correlation function to adapt to any changing relationship between the chest and tumor motion. The technical details of the Synchrony correlation/prediction process are presented in Chapter 7.

The hybrid optical/x-ray tracking system has another advantage over direct imaging for real-time motion tracking. In a real-time tracking system, the time delay (latency) between target

(a)

(b)

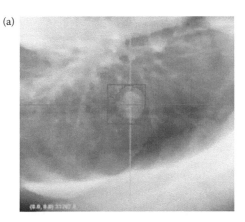 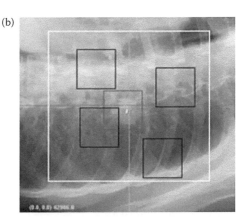

FIGURE 4.2 An illustration of the template-matching procedure used to detect lung tumors. (a) Shows the digitally reconstructed radiograph (DRR) from which the tumor template (box) is obtained; (b) illustrates how the template is scanned across a region of interest in an acquired radiograph. The correlation coefficient of the template with the underlying image reaches a maximum when the template is directly over the tumor.

detection and beam alignment response must be corrected by a predictive control loop algorithm in order that the beam and the tumor arrive simultaneously at the same position. The error in respiratory prediction algorithms generally tends to increase with increasing time delay (Murphy and Dieterich 2006). If an x-ray system is the sole source of target position data, then all of the time that is spent acquiring and analyzing the images contributes to the total latency time of the tracking response. This penalizes lengthy image analysis procedures that might otherwise provide a more accurate measure of tumor position. In the hybrid system, the robot responds directly to the optical tracking data. Thus, the system latency is determined by the speed of the optical tracker, which is in the order of 30 Hz. The x-ray images are used in parallel to maintain the tumor/chest correlation function, which does not need to be updated in real time. Thus, the x-ray image analysis process does not contribute to the system latency and can use as much time as is necessary to get an accurate and robust measure of the tumor's actual position.

Lung tumors can be difficult to discern in kV projection radiographs. If it is necessary to locate them automatically via a real-time image processing operation, then the problem becomes even more difficult. For this reason, the CyberKnife x-ray imaging system initially relied upon fiducials implanted in the lung tumor to detect its position in the radiographs. However, fiducial placement requires either bronchoscopic access to the tumor, or puncturing the lung one or more times with a needle, which is accompanied by the risk of pneumothorax. It is therefore desirable to avoid the use of fiducials whenever possible. The CyberKnife Synchrony system has recently been augmented with the Xsight® Lung Tracking System to allow direct tumor localization in certain instances (Fu et al 2007). This is done using the technique of template matching, in which a small image of the tumor shape is convolved over the x-ray image. Figure 4.2 illustrates the basic concept. A digitally reconstructed radiograph (DRR) of the vicinity of the tumor is computed from the treatment planning computed tomography study (Figure 4.2a). The DRR is filtered to suppress stationary features such as bony landmarks in order to enhance the tumor visibility. The immediate neighborhood of the tumor (the box in Figure 4.2a) is extracted as the template. The template is then scanned over a region of interest in each acquired radiograph (Figure 4.2b) while computing the correlation coefficient between the template and the underlying image content. When the template passes directly over the tumor, the correlation coefficient reaches a maximum, thus signaling the tumor's position. Fiducial-less lung tumor localization is presently viable for selected cases involving peripheral and apex tumors larger than about 15 mm in diameter (Fu et al 2007).

References

Adler JR, Murphy MJ, Chang S, and Hancock S, Image-guided robotic radiosurgery, *Neurosurgery*, 44, 1299–307, 1999.

Friedland JL, Freeman DE, Masterson-McGary ME, and Spellberg DM, Stereotactic body radiotherapy: An emerging treatment approach for localized prostate cancer, *Technol Cancer Res Treat*, 8(5), 387–92, 2009.

Fu D, Kahn R, Wang B, Wang H, Mu Z, Park J, Kuduvalli G, and Maurer CR, Xsight lung tracking system: A fiducial-less method for respiratory motion tracking, in *Robotic Radiosurgery: Treating Tumors That Move with Respiration*, HC Urschel Jr, ed., Springer-Verlag, Berlin, 2007.

Ho AK, Fu D, Cotrutz C, Hancock SL, Chang SD, Gibbs IC, Maurer CR, and Adler JR, A study of the accuracy of CyberKnife spinal radiosurgery using skeletal structure tracking, *Neurosurgery*, 60(2), 147–56, 2007.

King CR, Brooks JD, Gill H, Pawlicki T, Cotrutz C, and Presti JC, Stereotactic body radiotherapy for localized prostate cancer: interim results of a prospective phase II clinical trial, *Int J Radiat Oncol Biol Phys*, 73(4), 1043–48, 2009.

Murphy MJ, Tracking moving organs in real time, *Seminars in Radiation Oncology*, Chen and Bortfield editors, 14(1), 91–100, 2004.

Murphy MJ, Adler JR, Bodduluri M, Dooley J, Forster K, Hai J, Le Q-T, Luxton G, Martin D, and Poen J, Image-guided radiosurgery for the spine and pancreas, *Comp Aid Surg*, 5, 278–88, 2000.

Murphy MJ, Intra-fraction geometric uncertainties in frameless image-guided radiosurgery, *Int J Rad Onc Biol Phys*, 73(5), 1364–68, 2009.

Murphy MJ, Chang S, Gibbs I, Le Q-T, Martin D, and Kim D, Image-guided radiosurgery in the treatment of spinal metastases, *Neurosurg Focus*, 11(6), 1–7, 2001.

Murphy MJ, Chang S, Gibbs I, Le Q-T, Hai J, Kim D, Martin D, and Adler JR, Patterns of patient movement during image-guided frameless radiosurgery, *Int J Radiat Oncol Biol Phys*, 55(5), 1400–08, 2003.

Murphy MJ and Dieterich S, Comparative performance of linear and nonlinear neural networks to predict irregular breathing, *Phys Med Biol*, 51, 5903–14, 2006.

Ryu S, Kim D, Murphy MJ, Chang S, Le Q-T, Martin D, and Adler J, Image-guided frameless robotic stereotactic radiosurgery to spinal lesions, *Neurosurgery*, 49, 838–47, 2001.

Schweikard A, Glosser G, Bodduluri M, Murphy MJ, and Adler JR, Robotic motion compensation for respiratory movement during radiosurgery, *Comp Aid Surg*, 5, 263–77, 2000.

Shirato H, Shimizu S, Kunieda T, et al., Physical aspects of a real-time tumor-tracking system for gated radiotherapy, *Int J Rad Onc Biol Phys*, 48, 1187–95, 2000.

Van Herk M, Bemeijer P, and Lebesque JV, Inclusion of geometric uncertainties in treatment plan evaluation, *Int J rad Oncol Biol Phys*, 52(5), 1407–22, 2002.

Whyte RI, Crownover R, Murphy MJ, Martin DP, Rice TW, DeCamp MM, Rodebaugh R, Weinhous MS, and Le Q-T, Stereotactic radiosurgery for lung tumors: preliminary report of a phase I trial, *Ann Thorac Surg*, 75, 1097–101, 2003.

Fundamentals of Tracking with a Linac Multileaf Collimator

5.1   Introduction 33
5.2   Intrafraction Breathing Motion 34
5.3   The Early Work on Tracking One-Dimensional Motion 34
5.4   The Fatal Flaw of Motion Deconvolution Attempts 35
5.5   Tracking Two-Dimensional Motion 35
5.6   Tracking Delivery Design by Direct Aperture Optimization in 4D 35
5.7   Adaptive Therapy..36
5.8   Conclusions...36
Acknowledgments...36
References..37

Dualta McQuaid

Steve Webb

5.1 Introduction

In this chapter, firstly, the key steps of using external-beam radiation in the history of radiotherapy are briefly reviewed. Then follows some review of the types of intrafraction motion that are clinically observed, and then we develop the fundamentals of compensating for such intrafraction motion using a multileaf collimator that dynamically reshapes the treatment beam during each fraction. Methods to do this for regular motion in one and two dimensions are reviewed. These are the techniques that have been worked out in our centers. Glancing reference is made to the work of others in this area that is covered in other chapters in this book. The role of the alternative adaptive treatment is also discussed.

Radiotherapy for attempted cure of cancer is now more than 100 years old, having been first performed in 1896, a year after the discovery of the x-ray. For about 75 years, radiotherapy, in many complex guises, has been directed at the patient's tumor, guided only by plane x-radiographs and classical tomography. This situation changed in 1972 when commercial x-ray computed tomography (CT) became available; this technique was applied to the planning of external-beam radiotherapy. Shortly thereafter, tomographic MRI joined the image-guiding armamentarium and now these two imaging modalities are standard tools in the radiotherapy workup. Imaging tissue function through emission computed tomography (SPECT and PET) is also beginning to play a part. Whilst images have always guided radiotherapy, the science of image-guided radiation therapy is now regarded as one of the principal foci of the field (Bortfeld et al 2006; Timmerman and Xing 2009).

Tomographic imaging revolutionized the practice of radiotherapy, and new conformal radiotherapy (CFRT) techniques were developed to take advantage of this knowledge. Starting seriously in 1988, a specific form of CFRT—intensity-modulated radiation therapy (IMRT)—began to be developed with the goal of improving conformity particularly to planning target volumes (PTVs) with concave outlines in which resided organs at risk (Webb 2003, 2009). By the turn of the 21st century commercial products existed for planning and delivering IMRT and gradually clinical implementation began to escalate. Today IMRT is regarded as an absolute necessity for the treatment of certain tumors with minimum collateral damage to organs at risk (Palta and Mackie 2003; Schlegel and Mahr 2007; Webb 1993, 1997, 2000, 2004; Van Dyk 1999).

It might be reasonably argued that, provided the PTV is clearly established through such imaging techniques (and it is recognized that there are still significant unknowns in terms of tumor and normal-tissue function), the delivery of CFRT and specifically IMRT to *static* targets can be understood well from both a planning and delivery viewpoint. If the target does not change day to day, if the patient is set up the same way each day, and if the target does not move during the treatment, tailoring a high-dose volume to the PTV will be at the limit of what is possible, subject to the laws of the physics of photon–tissue interactions. None of these three provisos are, however, strictly possible. Dealing with the first two problems is a field outside the scope of this book [see review by Webb (2006c)]. Dealing with the last of them, the so-called intrafraction motion, is the subject of this book. In this chapter, we shall introduce the problem and some techniques for solving it using the tracking of the multileaf collimator (MLC) attached to the linac.

5.2 Intrafraction Breathing Motion

The action of respiration is controlled by the diaphragm and the intercostal muscles in the rib cage. During inhalation, the diaphragm contracts and moves down and the rib cage moves up and out. Normal exhalation is a passive action where the diaphragm and the intercostal muscles relax, causing the chest to elastically contract. In quiet breathing, where only a fraction of the total breathing capacity is used, the action of the diaphragm is dominant, and the cycle is typically not symmetric; the lungs rest at the full-exhale position longer than at the full inhale position. This observation lead to the approximation of the respiratory waveform by a sinusoid raised to an even power (Lujan et al 1999). Given the asymmetry observed in the respiratory cycle, it is perhaps not surprising that hysteresis was observed in the motion of many points in the lung, including lung tumors. The hysteresis in this case refers to the difference in the trajectory of a point on inhalation and exhalation. Different trajectories on inhalation and exhalation can be modeled by adjusting a parameter controlling the phase difference between motion in the different orthogonal directions.

The simple models of respiratory motion described can approximate resting periodic breathing patterns, but respiratory motion can be considerably more complex. Extreme changes to the respiratory cycle can result from sighs, coughs, or sneezes, but even within normal cycles, considerable variation can exist. Breathing can be predominantly due to diaphragm motion or chest motion or a mixture of both, and the mixture can fluctuate with time or with the patient's position or well being. Intercycle variability in breathing was assessed for normal volunteers by Tobin et al (1988), who parametrized the breathing cycles by tidal volume, breathing frequency, and other associated parameters, which were found to vary considerably in the short term between cycles and in the long term between days. Furthermore, the variations were more in older patients, which is important in radiotherapy as older people make up a large proportion of the patient population treated for lung cancer. Variations in the respiratory cycles including changes to breathing period, amplitude, and baseline drifts were observed (Seppenwoolde et al 2002) in patients undergoing gated radiotherapy treatments.

5.3 The Early Work on Tracking One-Dimensional Motion

The first attempts to consider intrafraction motion and to correct for it were based on the notion that such motion was (1) regular, (2) rigid-body, (3) one-dimensional (1D), and (4) along the direction of the MLC leaves. None of these four assumptions is correct; none precisely matches reality. However, they are first-order approximations, and a start has to be made somewhere. If those four assumptions do hold, then there follows a relatively simple concept: provided the radiation beam at any instant of time, t, can be matched to the instantaneous position of the PTV at that time, then the PTV becomes stationary in the moving frame of the MLC. The concept is simple. If the position, x, along the leaf path (leading or trailing) is x_0 at time t, when the target is assumed static, then that leaf must be moved to position $x(t)$ at time t. On the familiar "position (x)–time (t) diagram" of the dynamic MLC (dMLC) technique (Stein et al 1994, Svensson et al 1994 and Spirou and Chui 1994) each of the static coordinates (x_0,t) is shifted to $(x(t),t)$. This is done (differently) for every leaf pair and for every time. This was first proposed by Keall et al (2001) and further developed by Webb (2005a). The only proviso is that, if the MLC leaf trajectories are determined such that one of each pair is always travelling at maximum physical speed $\hat{w}$, then the addition of a compensating extra speed would lead to an unphysical situation. Consequently, the "breathing leaves method" requires the initial downregulation of $\hat{w}$ to a value such that no leaf-speed violations ever occur after the frame transformation. This is a standard feature of all further developments, both 1D and 2D, of this method.

For this to work, albeit accepting the four approximating assumptions, the motion of the leaves has to be precisely "phase locked" to the corresponding PTV motion. Webb (2005a) gave a specific geometrical construction, christened the Boyer–Strait–Webb (BSW) construction because it was a development of that of Boyer and Strait (1997) for static-target dMLC IMRT. Interestingly, the use of the BSW construction led to a demonstration of the partial success of the outcome of "near phase matching" of the PTV and the MLC. Provided the phase difference was not too great, a reasonable delivery still resulted. *En passant* we may say that when the leaves are set to not breath and the reconstruction is averaged over a random set of MLC phase breathing then the outcome is simply the static intensity distribution convolved with the probability density function of the PTV motion. That is a very powerful observation because it proved for the first time that the dMLC acts exactly like a compensator when the delivery is averaged over a full set of fractions. There is no interplay effect. This complements theoretical calculations by Bortfeld et al (2002).

Webb (2006a) went further to show that the effect of changed density during intrafraction breathing was quite small. Webb (2006b) quantified the effects of latency between measuring motion and implementing the correction. Webb and Binnie (2006) made a first stab at allowing for differential elastic (non-rigid body) intrafraction motion and showed that the optimum solution was to track the mean motion. This was followed up by McClelland et al (2007) with a specific implementation based on a motion model developed from 4D CT scanning.

A large body of theoretical work on 1D (and later 2D) tracking has also been performed at several other centers, notable among which are Virginia Commonwealth University and Stanford (Keall et al 2001, 2005, 2006; Sawant et al 2008) and Indiana University (Papiez 2003, 2004; Papiez and Rangaraj 2005; Rangaraj and Papiez 2005; McMahon et al 2007; and Papiez et al 2005, 2007). These topics are discussed by these authors in other chapters of this book.

5.4 The Fatal Flaw of Motion Deconvolution Attempts

If the rigid-body motion were 1D, regular and in the direction of the MLC leaves, then it might be thought that a solution to the problem would be to deconvolve this motion from the fluence calculated for a static target and deliver the outcome. It was initially an attractive idea, but it fatally flawed. When such deconvolutions take place, the outcome is the requirement to deliver a fluence distribution that includes regions of unphysical negative fluence (Webb 2005b). This naturally leads to the idea that one might constrain such a deconvolution to be positive and inspect whether the outcome is "good enough." This also turns out to be problematic because very high fluence peaks are generated. They could be tolerated if it could be guaranteed that the motion were as deconvolved. However, that cannot be the case for certain, and hence, such peaks would be dangerous to implement. This led to the concept of studying constrained deconvolution in which positivity and capped peaks were part of the methodology. This does achieve some success but is not the way to go forward (Webb 2007).

5.5 Tracking Two-Dimensional Motion

The hysteresis observed in the motion of tumors under respiration (Section 5.2) often means that the tumor trajectory cannot be adequately described by a linear trajectory. The curved target trajectory in the beam's eye view causes the target to move perpendicularly to the direction of leaf travel moving from one leaf pair to another. To successfully track targets such as this, the 1D tracking proposals outlined in Section 5.3 must be adapted to allow the tracking trajectory of one leaf pair to be transferred to next pair. This necessitates synchronization of the planned trajectories of leaf pairs across the leaf bank and an adaption strategy of the leaf trajectories in response to a motion perpendicular to the direction of leaf travel. Two such synchronization strategies were proposed: (1) a mid-time synchronization technique (Rangaraj and Papiez 2005), and (2) a dedicated strict full synchronization strategy (McQuaid and Webb 2006).

If the leaves travel along the x-axis across bixel positions x_i and different leaf pairs are fixed at positions along the y-axis, then by the mid-time synchronization technique, the mean arrival time of the leading and trailing leaves for each leaf pair is set equal at each bixel position x_i. In the full synchronization technique, the spatial distance between adjacent leaves in a leaf bank is constrained to be less than some tolerated distance. This is accomplished by adding factors to each bixel position by which both the leading and trailing leaves are delayed. Because of the complex interplay of the trajectories of different leading and trailing leaf pairs, this problem must be solved iteratively, whereas the mid-time strategy can be solved in a single step but cannot guarantee any set distance between the adjacent leaves. The mid-time synchronization strategy was tested with a discrete adaption technique, in which no change in leaf trajectories is applied until the target motion in the y-direction exceeds one half of the leaf width at this point the leaves travel at maximum velocity to take up the planned position of their neighbors. The full synchronization strategy was tested with a linear interpolating adaption strategy where a motion of the target in the y-direction causes the leaves to move to a trajectory formed by the linear interpolation of a leaf pair with its neighbor. Under tests, both techniques have shown to deliver perfect results with infinitesimal leaf widths and infinite maximum leaf velocity, but in practical situations and with relevant delivery constraints, both techniques suffer from delivery errors caused by sampling errors induced by the leaf width and the finite response time for the case of the discrete adaption technique and interpolation error in the case of the linear-interpolation technique. These error sources prevent perfect delivery, but the delivery errors in the presence of significant 2D motion have been demonstrated to be significantly better than no correction for both correction strategies. This improvement comes at the price of an increase in delivery time caused by leaf synchronization, but this increase in time has been shown to be relatively modest.

5.6 Tracking Delivery Design by Direct Aperture Optimization in 4D

Many of the issues with target-tracking strategies arise from planning the treatment on a static patient geometry and subsequently adapting or modifying the delivery to compensate for the target and normal-tissue motion. Many of the issues such as nonuniform displacements along a ray path or changes in equivalent-path length to the target are impossible to fully correct by the plan and adapt strategies. Further, adherence to delivery-system constraints enforces compromises, which result in a less-than-perfect delivery but these compromises such as those involved when the target moves in 2D in the beam's eye view are known to depend on the nature of the original plan. Some beam-fluence maps suffer a significantly greater error than others do. An alternative planning methodology would be to avoid these issues by including the tissue motion and tracking leaf motion in the optimization.

McQuaid and Webb (2008) designed such a system in which the delivery was broken down into a series of sequentially delivered tracking segments. Each tracking segment was in turn subdivided into a series of control points matched to sequential instances of tissue geometry from a 4D patient model. The distance of leaf motion between successive control points was constrained such that the mechanical constraints of leaf motion were maintained. The technique was tested with a simple plan with three beam angles and three tracking segments per angle for a 4D patient model and for a digital phantom model. In these tests, the degradation of the plan quality caused by tissue motion was shown to be restored to the plan quality equivalent to the static case. In delivery tests (see Figure 5.1), with an Elekta MLCi collimator design, a 4D tracking plan was delivered successfully to a phantom moving in an elliptical path in the beam's eye view (McQuaid et al. 2009). The completion of the delivery without interruption proves that the planning technique is capable of designing plans conforming to a complete set

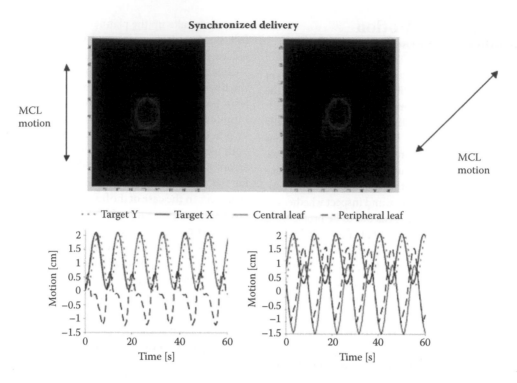

FIGURE 5.1 Synchronized leaf tracking performed on an Elekta linac. (Adapted from McQuaid, D., Partridge, M., Tayler, R., Evans, P.M., and Webb, S., *Phys. Med. Biol.* 54, 3563–3578, 2009.)

of delivery constraints, including the combination of leaf velocity constraints and minimum-gap constraints between directly opposing leaves and opposing leaves in adjacent tracks.

5.7 Adaptive Therapy

An alternative to tracking intrafraction motion is adaptive therapy. In this technique, the patient is allowed to breathe normally, and the breathing pattern is measured in some way. Armed with this knowledge, it is possible to compute the fluence that was really delivered at any given fraction. By comparing this with the prescribed fluence the difference can be established. Adaptive therapy aims to compensate for this difference by delivering a different fluence pattern at the subsequent fraction. This philosophy is carried right through a large number (maybe 30) of such fractions so that at the close, the residual error is small. The problem of augmenting the delivery by some kind of "breathing apparatus" is transformed into the problem of assessing the real motion and calculating its effect. Webb (2008) showed how this could be done by a methodology in fluence space, and Webb and Bortfeld (2008) showed an extension of this whereby dose space residuals were transformed into fluence-space changes for the subsequent fractions.

5.8 Conclusions

The current theory and methodology for delivering tracking radiation therapy to a target with intrafraction motion are

exceptionally well developed in stark contrast to the almost complete absence of clinical implementation so far. That is not surprising because to deliver safely tracking therapy, the engineering solution of many interconnected problems is required. Firstly, the actual, rather than the expected, motion must be measured and tracked. If that measurement is from some external sensor, then that has to be related to the internal tumor motion. The latency between making a measurement and using it for compensation has to be accommodated. Then the linac MLC itself requires to have fully integrated dynamic control linked to this measurement of tumor motion. Solutions are required to the problem of variable motion and elastic motion. So far, the only reported implementations have been experimental demonstrations of proof of principle using motion platforms (McQuaid et al 2009; Keall et al 2006; Sawant et al 2008). It is, however, known that all the three major linac/MLC manufacturers are in a race to provide a good commercial solution with this being sometimes in connection with VMAT/RapidArc dMLC delivery (McQuaid et al 2009; Court et al 2010; Falk et al 2010; Poulson et al 2010; Tacke et al 2010; Keall et al 2010; Krauss et al 2010).

Acknowledgments

The ICR/RMNHSFT is funded by Cancer Research, UK, and the National Institute for Health Research. The 2D tracking and 4D planning work was done when the first author was at ICR/RMNHSFT.

References

Bortfeld T, Jokivarsi K, Goitein M, and Jiang S (2002). Effects of intra-fraction motion on IMRT delivery: Statistical analysis and simulation. *Phys. Med. Biol.*. 47, 2203–2220.

Bortfeld T, Schmidt-Ullrich R, de Neve W and Wazer D (2006). *Image Guided IMRT*. Springer, Heidelberg, Berlin.

Boyer AL and Strait JP (1997). Delivery of intensity-modulated treatments with dynamic multileaf collimator. *Proceedings in 12th International Conference on the Use of Computers in Radiation Therapy (Salt Lake City, Utah, May 1997)*. Edited by DD Leavitt and G Starkschall, Medical Physics Publishing, Madison, WI, pp. 13–16.

Court L, Wagar M, Berbeco R, Reisner A, Winey B, Schofield D, Ionascu D, Allen A M, Popple R and Lingos T (2010). Evaluation of the interplay effect when using RapidArc to treat targets moving in the craniaocaudal or right-left direction. *Med. Phys.* 37, 4–11.

Falk M, Munck af Rosenchöld P, Keall P, Catell H, Cho B C, Poulson P, Povsner S, Sawant A, Zimmerman J, and Korreman S (2010). Real-time dynamic MLC tracking for inversely optimised arc therapy. *Radiother. Oncol.* 94, 218–223.

Keall PJ, Joshi S, Vedam SS, Siebers JV, Kini VR, and Mohan R (2005). Four-dimensional radiotherapy planning for DMLC-based respiratory motion tracking. *Med. Phys.* 32, 942–951.

Keall P, Sawant A, Cho B, Ruan D, Wu J, Poulsen P, Petersen J, Newell LJ, Catell H, and Korreman S (2010). Electromagnetic-guided dynamic multileaf collimator tracking enables motion management for intensity-modulated arc therapy. *Int. J. Rad. Oncol. Biol. Phys.* 79, 312–320.

Keall PJ, Siebers JV, Vedam S. and Mohan R (2001). Motion adaptive x-ray therapy: a feasibility study. *Phys. Med. Biol.* 46, 1–10.

Keall P, Vedam S, George R, Bartee C, Siebers J, Lerma F, Weiss E, and Chung T (2006). The clinical implementation of respiratory-gated intensity-modulated radiotherapy. *Med. Dosim.* 31, 152–162.

Krauss A, Nill S, Tacke M, and Oelfke U (2010). Electromagnetic real-time tumor position monitoring and dynamic multileaf collimator tracking using Siemens 160 MLC: geometric and dosimetric accuracy of an integrated system. *Int. J. Rad. Oncol. Biol. Phys.* 79, 579–587.

Lujan AE, Larsen EW, Balter JM, and Haken RKT (1999). A method for incorporating organ motion due to breathing into 3D dose calculations. *Med. Phys.* 26, 715–720.

McClelland J, Webb S, McQuaid D, Binnie DM, and Hawkes DJ (2007). Tracking "differential organ motion" with a "breathing" multileaf collimator: magnitude of problem assessed using 4D CT data and a motion-compensation strategy. *Phys. Med. Biol.* 52, 4805–4826.

McMahon R, Papiez L, and Rangaraj D (2007). Dynamic-MLC leaf control utilizing on-flight intensity calculations: a robust method for real-time IMRT delivery over moving rigid targets. *Med. Phys.* 34, 3211–3223.

McQuaid D, Partridge M, Symonds Tayler R, Evans P M, and Webb S (2009). Target-tracking deliveries on an Elekta linac: a feasibility study. *Phys. Med. Biol.* 54, 3563–3578.

McQuaid D and Webb S (2006). IMRT delivery to a moving target by dynamic MLC tracking: delivery for targets moving in two dimensions in the beam's-eye view. *Phys. Med. Biol.* 51, 4819–4839.

McQuaid D and Webb S (2008). Target-tracking deliveries using conventional multileaf collimators planned with 4D direct-aperture optimisation. *Phys. Med. Biol.* 53, 4013–4029.

Palta JR and Mackie TR (2003). *IMRT – the State of the Art*. Medical Physics Publishing, Madison, WI.

Papiez L (2003). The leaf sweep algorithm for an immobile and moving target as an optimal control problem in radiotherapy delivery. *Math. Comput. Model.* 37, 735–745.

Papiez L (2004). DMLC leaf-pair optimal control of IMRT delivery for a moving rigid target. *Med. Phys.* 31, 2742–2754.

Papiez L and Rangaraj D, (2005). DMLC leaf-pair optimal control for mobile, deforming target. *Med. Phys.*, 32, 275–285.

Papiez L, Rangaraj D, and Keall P (2005). Real-time DMLC IMRT delivery for mobile and deforming targets. Med. Phys. 32, 3037–3048.

Papiez L, McMahon R, and Timmerman R (2007). 4D DMLC leaf sequencing to minimize organ at risk dose in moving anatomy. *Med. Phys.* 34, 4952–4956.

Poulson PR, Cho B, Sawant A, and Keall PJ (2010). Implementation of a new method for dynamic miltileaf collimator tracking of prostate motion in arc radiotherapy using a single kV imager. *Int. J. Rad. Oncol. Biol. Phys.* 76, 914–923.

Rangaraj D and Papiez L (2005). Synchronized delivery of DMLC intensity modulated radiation therapy for stationary and moving targets. *Med. Phys.* 32, 1802–1817.

Sawant A, Venkat R, Srivastava V, Carlson D, Povzner S, Cattell H, and Keall P (2008). Management of three-dimensional intrafraction motion through real-time DMLC tracking. *Med. Phys.* 35, 2050–2061.

Schlegel W and Mahr A (2007). *3D Conformal Radiation Therapy*, 2nd edition. Springer, Heidelberg.

Seppenwoolde Y, Shirato H, Kitamura K, Shimizu S, van Herk M, Lebesque JV, and Miyasaka K (2002). Precise and real-time measurement of 3D tumor motion in lung due to breathing and heartbeat, measured during radiotherapy. *Int. J. Radiat. Oncol. Biol. Phys.* 53, 822–834.

Stein J, Bortfeld T, Dörschel B, and Schlegel W (1994). Dynamic x-ray compensation for conformal radiotherapy by means of multileaf collimation. *Radiother. Oncol.* 32, 163–173.

Svensson R, Källman P, and Brahme A (1994). Analytic solution for the dynamic control of multileaf collimators. *Phys. Med. Biol.* 39, 37–61.

Spirou SV and Chui CS (1994). Generation of arbitrary intensity profiles by dynamic jaws or multileaf collimators. *Med. Phys.* 21, 1031–1041.

Tacke MB, Nill S, Krauss A, and Oelfke U (2010). Real-time tumor tracking: Automatic compensation of target motion using the Siemens 160 MLC. *Med. Phys.* 37, 753–761.

Timmermann R and Xing L (2009). *Image Guided and Adaptive Radiation Therapy*. Wolters Kluwer/Lippincott Williams and Wilkins, Philadelphia.

Tobin MJ, Mador MJ, Guenther SM, Lodato RF, and Sackner MA (1988). Variability of resting respiratory drive and timing in healthy subjects. *J. Appl. Physiol.* 65, 309–317.

Van Dyk J (1999). The modern technology of radiation oncology. Medical Physics Publishing, Madison, WI.

Webb S (1993). *The Physics of Three-Dimensional Radiation Therapy—Conformal Radiotherapy, Radiosurgery and Treatment Planning*. IOP Publishing, Bristol.

Webb S (1997). *The Physics of Conformal Radiotherapy—Advances in Technology*. IOP Publishing, Bristol.

Webb S (2000). *Intensity Modulated Radiation Therapy*. IOP Publishing, Bristol.

Webb S (2003). *Historical Perspective on IMRT in Intensity-Modulated Radiation Therapy: the State of the Art*. Edited by JR Palta and TR Mackie. Medical Physics Publishing, Madison, WI. pp. 1–23.

Webb S (2004). *Contemporary IMRT—Developing Physics and Clinical Implementation*. IOP Publishing, Bristol.

Webb S (2005a). The effect on IMRT conformality of elastic tissue movement and a practical suggestion for movement compensation via the modified dynamic multileaf collimator (dMLC) technique. *Phys. Med. Biol.* 50, 1163–1190.

Webb S (2005b). Limitations of a simple technique for movement compensation via movement-modified fluence profiles. *Phys. Med. Biol.* 50, N155–N161.

Webb S (2006a). Does elastic tissue intrafraction motion with density changes forbid motion-compensated radiotherapy? *Phys. Med. Biol.* 51, 1449–1462.

Webb S (2006b). Quantification of the fluence error in the motion-compensated dynamic MLC (DMLC) technique for delivering intensity-modulated radiotherapy (IMRT). *Phys. Med. Biol.* 51, L17–L21.

Webb S (2006c). Motion effects in (intensity-modulated) radiation therapy—A review *Phys. Med. Biol.* 51, R413–R425.

Webb S (2007). Intrafraction motion compensation by highly constrained iterative deconvolution of organ motion *Phys. Med. Biol.* 52, N309–N320.

Webb S (2008). Adapting IMRT delivery fraction-by-fraction to cater for variable intra-fraction motion. *Phys. Med. Biol.* 53, 1–21.

Webb S (2009). The 21st Birthday Party for intensity-modulated radiation therapy (IMRT); 21 years from 1988–2009; From concept to practical reality. *Proceedings in World Congress on Medical Physics*, Munich, Sept 2009 Proc IFMBE Proceedings 25/I, p 49–52.

Webb S and Binnie DM (2006). A strategy to minimize errors from differential intrafraction organ motion using a single configuration for a "breathing" multileaf collimator. *Phys. Med. Biol.* 51, 4517–4531.

Webb S and Bortfeld T (2008). A new way of adapting IMRT delivery fraction-by-fraction to cater for variable intrafraction motion. *Phys. Med. Biol.* 53, 5177–5191.

$$6$$

Couch-Based Target Alignment

6.1 Introduction 39
6.2 Couch Shifts for Patient Alignment...39
6.3 Transient and Low-Frequency Couch-Based Target Alignment 41
6.4 Dynamic Couch-Based Target Motion Compensation 41
6.5 Quality Assurance 43
6.6 The Future of Couch-Based Motion Correction in Radiation Therapy......................... 44
References... 45

Kathleen T. Malinowski

Warren D. D'Souza

6.1 Introduction

For many years, developments for target alignment in radiation therapy were focused on beam shaping and radiation source positioning technologies (Sawant 2008; Seppenwoolde 2007). Recent developments in treatment couch technology have led to a surge in couch-based target alignment methods (Figure 6.1). New treatment couches are fully automated, more accurate, faster, and, in some cases, capable of exhibiting more degrees of freedom in patient positioning than their predecessors. Such developments have expanded the role of the treatment couch in radiation therapy.

The treatment couch is an important component of a clinical radiation therapy treatment system. An indexed treatment couch is needed for precise patient positioning and is often guided by external fiducial marks or in-room imaging systems. Some couches are capable of rotation, thus enabling corrections in target orientation. Given the growing level of couch technology sophistication, the treatment couch's exact position can be related to the coordinate system of the patient anatomy. Modern record and verify systems make it possible to detect certain patient setup errors by comparing the couch position on a particular day to previous couch positions recorded over the course of treatment. Treatment couch automation has decreased patient setup time and increased the accuracy of target alignment. Recently, treatment couch systems that can correct for intrafraction tumor motion have begun to emerge.

In this chapter, we describe the role of the treatment couch in patient alignment. We focus on three categories of target alignment: (1) shifts to correct for patient setup prior to the start of treatment (or sometimes referred to as interfraction setup errors), (2) shifts to correct for transient changes during treatment (such as the passage of bowel gas and peristalsis) or changes that manifest over a longer period (>6 seconds) (such as anatomic position drifts), and (3) shifts *to* correct for changes

as they occur in real time (such as respiration-induced target motion). We also briefly discuss quality assurance associated specifically with couch-based patient alignment and the future of couch-based motion correction.

6.2 Couch Shifts for Patient Alignment

Positioning the patient in order to maintain proper alignment of the tumor with the radiation beam is an integral part of the patient setup procedure for radiation therapy. Generally, the couch is used for initial patient alignment with skin-based setup marks after the patient assumes a prone or supine position. Often, the target is then localized for more precise couch positioning using devices such as kilovoltage- or megavoltage-energy x-ray imaging systems (Mutanga 2008), cone-beam computed tomography (CBCT) (Guan 2009), electromagnetic implanted fiducial marker localization (Kupelian 2007), ultrasound (Foster 2010), or optical surface imaging (Schöffel 2007). Chapter 1 has already presented an overview of techniques for localizing the target in the treatment room. This chapter will focus on the use of the couch for target repositioning once these measurements have been made.

Couch positioning for target alignment is becoming an increasingly automated process. Until recently, repositioning the couch required manually moving the couch into position. Most modern treatment couches incorporate a controller that is capable of automatically moving the treatment couch the correct distance to a specified position. Brock et al. (2002) and Woo et al. (2002) argue that remote couch positioning is more accurate than manual positioning because it reduces the chance of human error. Brock et al. (2002) also have shown that automated couch positioning systems can reposition the couch more quickly than manual couch positioning systems, even if the system must be activated from inside the treatment room. Still, because of the maze structure (distance the therapist would have to walk) and

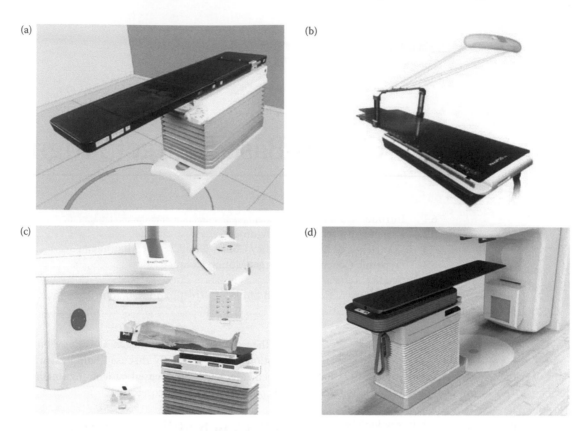

FIGURE 6.1 Technologically advanced treatment couch systems. (a) IGRT Couch Top System (Aktina Medical, Congers, NJ). (b) HexaPOD™ evo RT System (Elekta AB, Stockholm, Sweden). (c) Exactrac® Robotics module (Brainlab AG, Feldkirchen, Germany). (d) Protura™ 6DOF Robotic Couch (CIVCO Medical Solutions, Kalona, Iowa).

door shielding (the speed of opening and closing the door is directly related to the weight of the door) of the treatment room, repositioning the couch from the treatment console can lead to a significant reduction in the time required to set up a patient. Malinowski et al. (2008) have shown that for optimal efficiency, clinical protocols for shifting the couch should differ depending on whether the couch can be repositioned remotely.

Most couch-based target alignment consists of translational shifts. However, for certain tumor sites, rotating the target to its planned orientation can provide significant improvements in dose delivery. Van Herten et al. (2008) have shown that rotation about the vertical axis (couch kick or yaw) can provide a modest improvement in target coverage of the prostate; on average, the coverage of the seminal vesicles improved from 92.6% to 95.9%. Redpath et al. (2008) concluded that translation is more important than rotation in aligning the prostate or bladder, but that rotation is important for some patients. Yue et al. (2006) have shown that dose to parts of a brain tumor in a head phantom can be improved by 8% through 6D (translation plus yaw, pitch, and roll) alignment. On the other hand, Wu et al. (2009) concluded that rotation and deformation of lung tumors were negligible in comparison to translational motion. Site-specific benefits of rotational correction may drive the development of treatment couch-based motion compensation systems for particular treatment sites.

Recently, platform systems capable of 6D positioning have been introduced for radiation therapy applications (Figure 6.1b–d). Meyer et al. (2007) have explored the use of a six-degree-of-freedom treatment couch for patient setup. Although the translational and rotational positioning accuracies of the system were better than 1 mm and 1°, many of the shifts needed could not be accomplished through motion of the 6D platform alone. A 3°-change in pitch, for instance, resulted in a 9-cm vertical shift of the isocenter, which was beyond the vertical translation limits of the platform. Because this 6D platform is attached to a conventional treatment couch positioning system, Wilbert et al. (2010b) have described a procedure in which the patient is manually shifted to the approximate setup position prior to the CBCT acquisition and subsequent 6D correction. Used in conjunction with optical tracking for the initial alignment, this procedure took an average of 4.5 minutes to complete.

Rotating the patient support system, however, may have unintended side effects. Van Herten et al. (2008) have shown that rotating the support system to align the target can lead to overdosing of distant organs at risk. In the worst case, dose to the femoral heads increased by 12%. In addition, Guckenberger et al. (2007) and Linthout et al. (2007) have investigated secondary motion of the patient anatomy occurring because of 6D alignment. Guckenberger et al. (2007) considered cranial, prostate, and other cases and reported a significant correlation

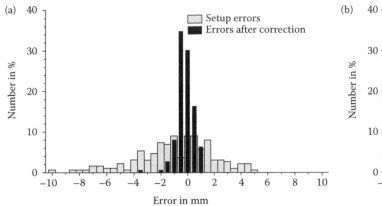
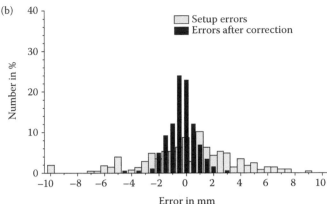

FIGURE 6.2 Distribution of setup errors and errors after 6 DOF correction for (a) immobilized (n = 65) and (b) nonimmobilized (n = 69) patients. (Reproduced from Guckenberger, M., Meyer, J., Wilbert, J., Baier, K., Sauer, O., and Flentje, M., *Strahlenther. Onkol.*, 183, 307–13, 2007. With permission.)

between rotation magnitude and setup errors. Rotation about the left–right (LR) axis leads to a shift in the cranial–caudal (CC) direction, and rotation about the CC axis leads to a shift in the LR direction. A rotation of 3° resulted in a 2-mm shift downward in the bony anatomy. Patient immobilization using the BodyFIX (Elekta, Stockholm, Sweden) system reduced these errors considerably (Figure 6.2). As a result, Guckenberger et al. recommended rotational error correction only with adequate patient fixation. In a study of prostate setup on a treatment couch capable of 6D correction, Linthout et al. found that 15% of patients show secondary motion greater than 2.0 mm or 1.0°, and 2% of patients show secondary motion greater than 3.0 mm or 2.0°. Most (92%) of the patients never exceeded the 3 mm/2.0° threshold. For the larger threshold, the subsequent secondary motion can be predicted from secondary motion measurements in the first five fractions with sensitivity of 80% and specificity of 97%.

Yue et al. (2008) have presented a strategy for circumventing the above consequences of rotating the patient support system by optimizing translational corrections to compensate for translational, rotational, and deformable target deviations. The CTV coverage and the minimum dose to the CTV could be improved significantly by translating the couch to optimize for volume or dose coverage. Yue et al. (2006) have also described a method of correcting for rotation and translation by translating the couch and altering gantry and collimator orientations exclusively. Similarly, Boswell et al. (2005) have described a method for slowly translating the treatment couch in the anterior–posterior (AP) or LR direction to correct for the pitch and yaw setup errors during helical tomotherapy treatments.

6.3 Transient and Low-Frequency Couch-Based Target Alignment

Recent developments in real-time target motion monitoring have made it possible to use the couch to intermittently realign the target over the course of treatment. The target shifts in such cases may be attributed to transient or low-frequency events such as the

passage of bowel gas, peristalsis, and anatomic drifts. The Calypso® System (Calypso Medical, Seattle, WA), which uses nonionizing electromagnetic radiation to track specialized implanted fiducial markers, can be used to monitor prostate position from the treatment console. Similarly, the AlignRT® system (Vision RT LTD, London, United Kingdom) monitors breast alignment through optical 3D surface tracking. Both the Calypso® system and the AlignRT® system can control certain couches by interfacing with them (e.g., the Varian 4D Integrated Treatment Console, 4DITC, Varian, Palo Alto, CA). When the target moves away from treatment isocenter, the therapist can interrupt the beam and use the 4DITC to reposition the patient through couch shifts calculated by the tracking devices. Following the repositioning of the target, the treatment can resume.

For safety purposes, most remote couch-shift systems are not fully automated. Instead, human intervention is required to initiate couch shifts. An exception has been described by Wiersma et al. (2010), who have developed a fully automated repositioning platform for the head that operates throughout the duration of treatment fraction. The system monitors head position through an optically tracked bite block device. When the target alignment error exceeds 0.2 mm, the system automatically translates the head in 3D to correct for the intrafraction motion. Using a head cradle alone for immobilization, this system can keep a target within 0.5 mm of isocenter 95% of the time (Figure 6.3).

6.4 Dynamic Couch-Based Target Motion Compensation

Currently, continuously moving couches are utilized in only a few radiation therapy imaging and treatment techniques, including CT imaging, helical tomotherapy, and translational total body irradiation. Recently, development has begun on dynamic couch systems that are designed to compensate for real-time intrafraction target motion. A continuous treatment couch motion compensation system could be integrated into a conventional treatment couch, which allows interaction with

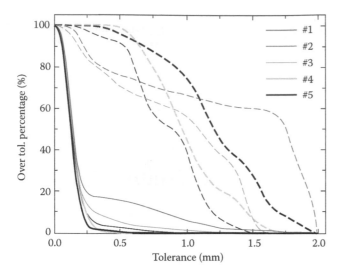

FIGURE 6.3 (**See color insert.**) Percentage of time that a particular tolerance was exceeded over a 15 min treatment period. Both corrected (*solid*) and uncorrected (*dashed*) motions are displayed for each test subject. (Reproduced from Wiersma, R.D., Wen, Z., Sadinski, M., Farrey, K., and Yenice, K.M. *Phys. Med. Biol.* 55, 389–401, 2010. With permission.)

other machine components such as the gantry or the multileaf collimator (MLC) leaves. As a result, couch-based motion compensation is emerging as a viable solution to be used in conjunction with more complex treatment-delivery modalities such as intensity-modulated radiation therapy, 3D conformal arc therapy, and intensity-modulated arc therapy.

D'Souza et al. (2005) first investigated the technical feasibility of using the treatment couch to compensate for real-time respiration-induced tumor motion using a miniaturized model of a treatment couch system. In this study, a platform simulated tumor motion in a measurement phantom while a second platform underneath the phantom platform simulated a treatment couch. The platforms were linked by an electronic loop that directed the two platforms to move in opposite directions in phase with one another. The couch-based phantom motion compensation was able to correct much of the dose blurring that occurred as a result of phantom motion (Figure 6.4). In a related study, Wilbert et al. (2008) reported on the performance of a robotic tabletop in compensating for elliptical motion in the coronal plane. Depending on the motion profile, this system reduced motion of the phantom with respect to the radiation source by 13%–50%.

In a feasibility study, Wilbert et al. (2008) highlighted the importance of hardware design in a motion-compensating couch system. In their experiments, the table under evaluation was unable to compensate for motion in which the velocity was greater than 1 cm/s. By contrast, Shirato et al. (2006) has measured tumor velocities of up to 9.4 cm/s. Acceleration limits and other aspects of system dynamics will also factor into the system's ability to fully compensate for tumor motion. D'Souza et al. (2006) theoretically analyzed a dynamic model for a couch system. In simulations exploring system latencies and dynamics

associated with a mechanical couch, they showed that the performance of a feedback control system was highly dependent upon both controller design and velocity and acceleration limits.

The multiple components in a treatment couch-based motion compensation system can be configured according to a number of possible control schemas. Careful selection of the controlled variable is necessary for system stability. Couch-based motion compensation systems are closed looped by nature because the very act of correcting for tumor motion moves the tumor in space. Still, interaction between respiratory disturbances and dynamics of the mechanical system could lead to runaway positive feedback. This condition can be avoided by using the tumor position with respect to the treatment couch surface as the controlled variable. Thoughtful control system design will be vital to the safety of these systems.

In general, target motion compensation errors can be minimized by designing a controller that inverts the dynamics of the couch system. However, system dead time, the time taken by a device to respond to a command, cannot be inverted. A couch system will also have some finite latency, which is a delay associated with communication and processing. The system described by Wilbert et al. (2008) had an image-based tumor tracking latency of 300–400 ms, a large delay in the context of typical breathing periods of 4–5 seconds. The standard method to deal with this latency is to predict the future position of the target (Murphy 2002). This is only feasible for systematic movements such as those associated with breathing. A more detailed discussion of the latency prediction problem is presented in Chapter 7.

Patient comfort is a key consideration in the clinical feasibility of a continuously moving treatment couch. In particular, the magnitudes, velocities, accelerations, and periodicity of respiration-induced motion have led to concerns for motion sickness. D'Souza et al. (2009) have investigated this issue with a validated 16-question motion sickness assessment questionnaire (Gianaros et al., 2001). Although 32% of the 50 patient-volunteers reported a history of motion sickness, none of the subjects experienced motion sickness during or after lying on couch moving through a realistic 2D tumor trajectory. Similarly, Sweeney et al. (2009) investigated the comfort of 23 patients and 10 healthy subjects who lay on a treatment couch that was moved through a 3D cyclical tumor path incorporating 2 degrees of rotational motion about the cranial–caudal axis. In this case, only one subject, who regularly took antinausea medication while traveling, reported slight discomfort.

Another challenge in the development of a motion-compensating treatment couch is the possibility that the continuous couch motion will induce anatomical changes. Two studies have looked for abnormal external tissue motion for patients on a moving couch. D'Souza et al. (2009) characterized the relative motion of markers affixed to different parts of the torso through regression analysis and phase offset measurements. No systematic change was seen between stationary and moving couch conditions, even in patients weighing more than 85 kg or with a weight–height ratio greater than 0.49 kg/cm. Wilbert et al. (2010a), however, did observe lateral oscillation of the abdominal wall when the couch was moved

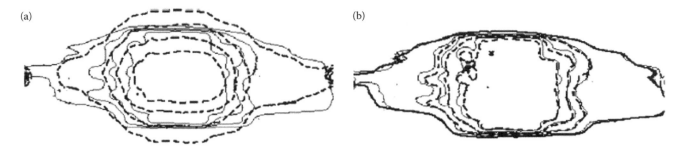

FIGURE 6.4 Isodose (90%, 70%, 50%, and 30% lines) comparison in the coronal plane for a step-and-shot plan delivered on a solid water phantom moving in the superior-inferior direction with an amplitude of 2 cm. Panel (a) represents static (*solid*) and phantom motion (*dashed*) cases, and panel (b) represents static (*solid*) and miniature couch-based phantom motion compensation (*dashed*) cases. (Reproduced from D'Souza, W.D., Naqvi, S.A., and Yu, C.X. *Phys. Med. Biol.* 50, 4021–33, 2005. With permission.)

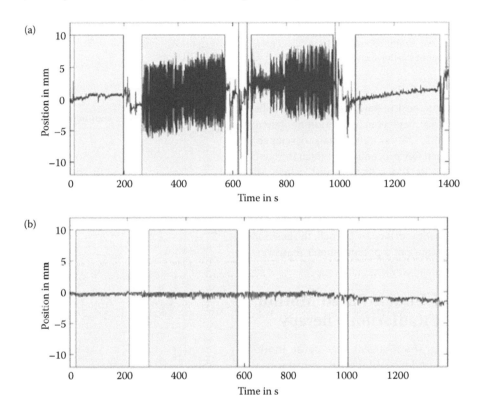

FIGURE 6.5 Lateral motion of the abdomen during measurements 1–4 (shaded boxes). In (a) the test person is placed on the table without fixation; in (b) the same test person is fixed in the BodyFIX system. Resonances during measurements with breathing-correlated table motion (Measurements 2 and 3) are completely eliminated by the BodyFIX system. (Reproduced from Wilbert, J., Baier, K., Richter, A., Herrmann, C., Ma, L., Flentje, M., and Guckenberger, M., *Int. J. Radiat. Oncol. Biol. Phys.* 77(2), 622–629, 2010a. With permission)

in either the CC or the AP direction. In one patient weighing 140 kg, the oscillation reached an amplitude of 10 mm (Figure 6.5). For this patient, the oscillations were eliminated by immobilization with the BodyFIX (Medical Intelligence) vacuum cushion system.

For respiration-induced tumor motion, a motion-compensating couch system will move in real time according to the patient's breathing, a physiological function that a patient can control consciously. Wilbert et al. (2010a) investigated whether breathing patterns change when a patient's breathing controls couch motion. Although changes in breathing patterns sometimes occurred during the experiment, no clear causal relationship could be established, and changes were generally small and gradual. Malinowski et al. (2011) have shown that changes in breathing patterns over the duration of a treatment fraction are normal, even without a motion-compensating treatment couch. Thus, there is no clear evidence that breathing-controlled couch motion leads to breathing pattern changes.

6.5 Quality Assurance

Remote positioning systems add complexity to radiation therapy devices, necessitating the development of novel quality

TABLE 6.1   Performance of Remote Treatment Couch Positioning Devices

| Authors | System | Accuracy | | | | | | | |
| | | LR (mm) | | CC (mm) | | AP (mm) | | Rotation (°) | |
		Mean	SD	Mean	SD	Mean	SD	Mean	SD
Bel et al.	Siemens ZXT	0.03	0.27	0.09	0.22	0.03	0.26	−0.00	0.04
Guan et al.[a]	Varian	<2	–	<2	–	<2	–	–	–
Li et al.	Elekta	0.16	0.48	0.11	0.12	0.32	0.30	–	–
Mutanga et al.	Elekta TCSA	0.29	0.27	0.03	0.21	0.23	0.28	–	–
Takakura et al.[b]	BrainLAB Robotic Tilt Module	<0.06	0.15	<0.06	0.12	<0.06	0.10	−0.05 0.02 −0.01	0.14 0.10 0.07
Van de Vondel et al.	Precitron Hercules	−0.10	0.30	0.00	0.30	0.70	0.53	0.21	0.23
Wilbert et al. (2008)[b]	Medical Intelligence HexaPOD	0.1	0.1	0.3	0.2	0.2	0.1	0.08 0.06 0.08	0.1 0.1 0.1

[a] A single set of translational error results representing LR, CC, and AP motion was reported.
[b] Means and standard deviations of rotational errors are in the order of yaw, pitch, and roll.

assurance (QA) procedures to ensure patients' safety. Published performances of remote treatment couch positioning systems are summarized in Table 6.1. Li et al. (2009) have developed a daily remote treatment couch QA procedure in which the couch is shifted remotely after CBCT acquisition and before MV orthogonal image acquisitions. The actual couch shift is measured through kV/MV image registration. In trials, this QA procedure was able to detect positioning errors greater than the tolerance of 2 mm. In the case described by Li et al., the positioning was out of tolerance because of a potentiometer (hardware) failure.

6.6 The Future of Couch-Based Motion Correction in Radiation Therapy

At the time of this writing, the first real-time couch motion compensation systems are under development. A major challenge in their implementation is the smooth integration of the couch itself (motors, motion controllers, and drivers) with such varied tumor localization components as x-ray imaging, optical or electromagnetic tracking, and computer processing systems. Although a few systems capable of real-time tumor localization are clinically available, development of safe and accurate tumor localization systems for a wide range of tumor sites continues to be an area of active research. The latencies associated with the mechanical system and communication between components will be important factors in overall system performance.

As real-time motion compensation devices are completed and are translated into the clinic, it is likely that these devices will be combined to best take advantage of the strengths of each system. Podder et al. (2008) have implemented a controller that splits tumor motion into high and low-frequency components (Figure 6.6) that are then compensated by the MLC leaves and

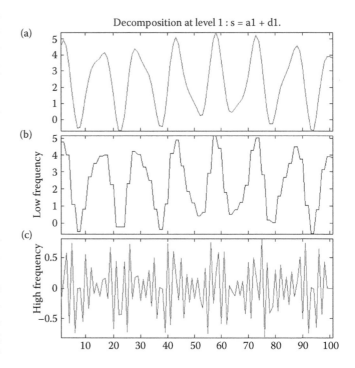

FIGURE 6.6 Wavelet decomposition of the tumor motion, (a) resultant motion, (b) low frequency component, and (c) high frequency component. (Reproduced from Podder, T.K., Buzurovic, I., Galvin, J.M., and Yu, Y., *IEEE Int. Conf. Robot. Autom.* 19, 2496–502, 2008. With permission.)

the couch, respectively. Such a configuration reduces the burden of high accelerations that must be accomplished by the couch while allowing for tumor motion that exceeds the displacement limitations of the MLC leaves. Future motion-compensation technologies are likely to be convergent.

References

Bel A, Petrascu O, Van de Vondel I, Coppens L, Linthout N, Verellen D, and Storme G. A computerized remote table control for fast on-line patient repositioning: Implementation and clinical feasibility. *Med. Phys.* 27(2):354–8, 2000.

Boswell SA, Jeraj R, Ruchala KJ, Olivera GH, Jaradat HA, James JA, Gutierrez A, Pearson D, Frank G, and Mackie TR. A novel method to correct for pitch and yaw patient setup errors in helical tomotherapy. *Med. Phys.* 32(6):1630–9, 2005.

Brock KK, McShan DL, and Balter JM. A comparison of computer-controlled versus manual on-line patient setup adjustment. *J. Appl. Clin. Med. Phys.* 3(3):241–7, 2002.

D'Souza WD, Naqvi SA, and Yu CX. Real-time intra-fraction-motion tracking using the treatment couch: a feasibility study. *Phys. Med. Biol.* 50(17):4021–33, 2005.

D'Souza WD, McAvoy TJ. An analysis of the treatment couch and control system dynamics for respiration-induced motion compensation. *Med. Phys.* 33(12):4701–9, 2006.

D'Souza WD, Malinowski KT, Van Liew S, D'Souza G, Asbury K, McAvoy TJ, Suntharalingam M, and Regine WF. Investigation of motion sickness and inertial stability on a moving couch for intra-fraction motion compensation. *Acta Oncol.* 48(8):1198–203, 2009.

Foster RD, Solberg TD, Li HS, Kerkhoff A, Enke CA, Willoughby TR, Kupelian PA. Comparison of transabdominal ultrasound and electromagnetic transponders for prostate localization. *J. Appl. Clin. Med. Phys.* 11(1):2924, 2010.

Gianaros PJ, Muth ER, Mordkoff JT, Levine ME, and Stern RM. A questionnaire for the assessment of the multiple dimensions of motion sickness. *Aviat. Space Environ. Med.* 72(2):115–9, 2001.

Guan H, Hammoud R, Yin FF. A positioning QA procedure for 2D/2D (kV/MV) and 3D/3D (CT/CBCT) image matching for radiotherapy patient setup. *J. Appl. Clin. Med. Phys.* 10(4):2954, 2009.

Guckenberger M, Meyer J, Wilbert J, Baier K, Sauer O, and Flentje M. Precision of image-guided radiotherapy (IGRT) in six degrees of freedom and limitations in clinical practice. *Strahlenther Onkol.* 183(6):307–13, 2007.

Kupelian P, Willoughby T, Mahadevan A, Djemil T, Weinstein G, Jani S, Enke C, Solberg T, Flores N, Liu D, Beyer D, and Levine L. Multi-institutional clinical experience with the Calypso System in localization and continuous, real-time monitoring of the prostate gland during external radiotherapy. *Int. J. Radiat. Oncol. Biol. Phys.* 67(4):1088–98, 2007.

Li W, Moseley DJ, Manfredi T, and Jaffray DA. Accuracy of automatic couch corrections with on-line volumetric imaging. *J. Appl. Clin. Med. Phys.* 10(4):3056, 2009.

Linthout N, Verellen D, Tournel K, Reynders T, Duchateau M, and Storme G. Assessment of secondary patient motion induced by automated couch movement during on-line 6 dimensional repositioning in prostate cancer treatment. *Radiother. Oncol.* 83(2):168–74, 2007.

Malinowski KT, Noel C, Roy M, Willoughby T, Djemi T, Jani S, Solberg T, Liu D, Levine L, and Parikh PJ. Efficient use of continuous, real-time prostate localization. *Phys. Med. Biol.* 53(18):4959–70, 2008.

Malinowski K, McAvoy TJ, George R, Dieterich S, D'Souza WD. Incidence of changes in respiration-induced tumor motion and its relationship with respiratory surrogates during individual treatment fractions. *Int. J. Radiat. Oncol. Biol. Phys.* 2011 Apr 16. [Epub ahead of print].

Meyer J, Wilbert J, Baier K, Guckenberger M, Richter A, Sauer O, and Flentje M. Positioning accuracy of cone-beam computed tomography in combination with a HexaPOD robot treatment table. *Int. J. Radiat. Oncol. Biol. Phys.* 67(4):1220–8, 2007.

Murphy MJ, Jalden J, and Isaksson M, Adaptive filtering to predict lung tumor breathing motion during image-guided radiation therapy, *Proceedings of the 16th International Congress on Computer-Assisted Radiology and Surgery*, Paris, 539–544, 2002.

Mutanga TF, de Boer HC, van der Wielen GJ, Wentzler D, Barnhoorn J, Incrocci L, and Heijmen BJ. Stereographic targeting in prostate radiotherapy: speed and precision by daily automatic positioning corrections using kilovoltage/megavoltage image pairs. *Int. J. Radiat. Oncol. Biol. Phys.* 71(4):1074–83, 2008.

Podder, TK, Buzurovic, I, Galvin, JM, and Yu, Y. Dynamics-based decentralized control of robotic couch and multi-leaf collimators for tracking tumor motion. *IEEE International Conference on Robotics and Automation.* 19(23):2496–502, 2008.

Redpath AT, Wright P, and Muren LP. The contribution of on-line correction for rotational organ motion in image-guided radiotherapy of the bladder and prostate. *Acta Oncol.* 47(7):1367–72, 2008.

Sawant A, Venkat R, Srivastava V, Carlson D, Povzner S, Cattell H, and Keall P. Management of three-dimensional intrafraction motion through real-time DMLC tracking. *Med. Phys.* 35(5):2050–61, 2008.

Schöffel PJ, Harms W, Sroka-Perez G, Schlegel W, and Karger CP. Accuracy of a commercial optical 3D surface imaging system for realignment of patients for radiotherapy of the thorax. *Phys. Med. Biol.* 52(13):3949–63, 2007.

Seppenwoolde Y, Berbeco RI, Nishioka S, Shirato H, Heijmen B. Accuracy of tumor motion compensation algorithm from a robotic respiratory tracking system: A simulation study. *Med. Phys.* 34(7), 2774–84, 2007.

Shirato H, Suzuki K, Sharp GC, Fujita K, Onimaru R, Fujino M, Kato N, Osaka Y, Kinoshita R, Taguchi H, Onodera S, and Miyasaka K. Speed and amplitude of lung tumor motion precisely detected in four-dimensional setup and in real-time tumor-tracking radiotherapy. *Int. J. Radiat. Oncol. Biol. Phys.* 64(4):1229–36, 2006.

Sweeney RA, Arnold W, Steixner E, Nevinny-Stickel M, and Lukas P. Compensating for tumor motion by a 6-degree-of-freedom treatment couch: Is patient tolerance an issue? *Int. J. Radiat. Oncol. Biol. Phys.* 74(1):168–71, 2009.

Takakura T, Mizowaki T, Nakata M, Yano S, Fujimoto T, Miyabe Y, Nakamura M, and Hiraoka M. The geometric accuracy of frameless stereotactic radiosurgery using a 6D robotic couch system. *Phys. Med. Biol.* 55(1):1–10, 2010.

Van de Vondel I, Coppens L, Verellen D, Linthout N, Van Acker S, and Storme G. Remote control for a stand-alone freely movable treatment couch with limitation system. *Med. Phys.* 28(12):2518–21, 2001.

van Herten YR, van de Kamer JB, van Wieringen N, Pieters BR, and Bel A. Dosimetric evaluation of prostate rotations and their correction by couch rotations. *Radiother. Oncol.* 88(1):156–62, 2008.

Wiersma RD, Wen Z, Sadinski M, Farrey K, and Yenice KM. Development of a frameless stereotactic radiosurgery system based on real-time 6D position monitoring and adaptive head motion compensation. *Phys. Med. Biol.* 55(2):389–401, 2010.

Wilbert J, Meyer J, Baier K, Guckenberger M, Herrmann C, Hess R, Janka C, Ma L, Mersebach T, Richter A, Roth M, Schilling K, and Flentje M. Tumor tracking and motion compensation with an adaptive tumor tracking system (ATTS): system description and prototype testing. *Med. Phys.* 35(9):3911–21, 2008.

Wilbert J, Baier K, Richter A, Herrmann C, Ma L, Flentje M, and Guckenberger M. Influence of continuous table motion on patient breathing patterns. *Int. J. Radiat. Oncol. Biol. Phys.* 77(2):622–9, 2010a.

Wilbert J, Guckenberger M, Polat B, Sauer O, Vogele M, Flentje M, and Sweeney RA. Semi-robotic 6 degree of freedom positioning for intracranial high precision radiotherapy; first phantom and clinical results. *Radiat. Oncol.* 5:42, 2010b.

Woo MK and Kim B. An investigation of the reproducibility and usefulness of automatic couch motion in complex radiation therapy techniques. *J. Appl. Clin. Med. Phys.* 3(1):46–50, 2002.

Wu J, Lei P, Shekhar R, Li H, Suntharalingam M, D'Souza WD. Do tumors in the lung deform during normal respiration? An image registration investigation. *Int. J. Radiat. Oncol. Biol. Phys.* 75(1):268–75, 2009.

Yue NJ, Knisely JP, Song H, and Nath R. A method to implement full six-degree target shift corrections for rigid body in image-guided radiotherapy. *Med. Phys.* 33(1):21–31, 2006.

Yue NJ, Kim S, Lewis BE, Jabbour S, Narra V, Goyal S, and Haffty BG. Optimization of couch translational corrections to compensate for rotational and deformable target deviations in image guided radiotherapy. *Med. Phys.* 35(10):4375–85, 2008.

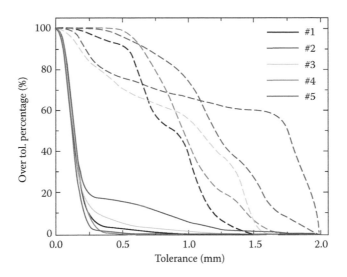

FIGURE 6.3 Percentage of time that a particular tolerance was exceeded over a 15 min treatment period. Both corrected (*solid*) and uncorrected (*dashed*) motions are displayed for each test subject. (Reproduced from Wiersma, R.D., Wen, Z., Sadinski, M., Farrey, K., and Yenice, K.M. *Phys. Med. Biol.* 55, 389–401, 2010. With permission.)

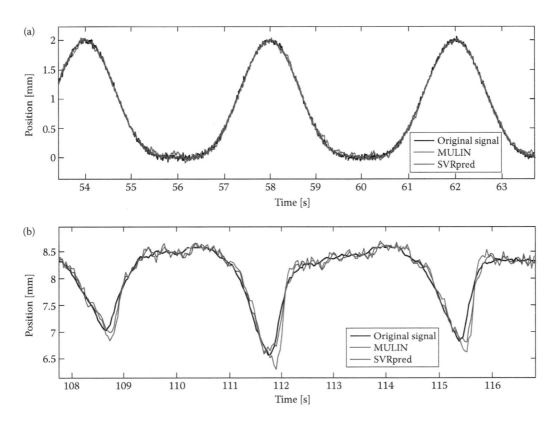

FIGURE 7.13 Results of the MULIN and SVRpred algorithms on the simulated signal with noise (a) and the real signal (b). The original signal is shown in black, the MULIN algorithm in blue and the SVRpred algorithm in red.

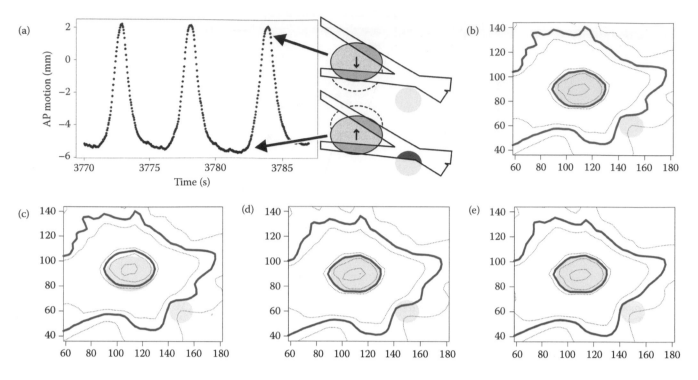

FIGURE 8.5 (a) a respiratory motion pattern and the position of beams and VOI during inhalation (planning state) and exhalation; (b–e) dose distributions after planning, for no motion compensation, for motion compensation with 3D planning, and for motion compensation with 4D planning, respectively.

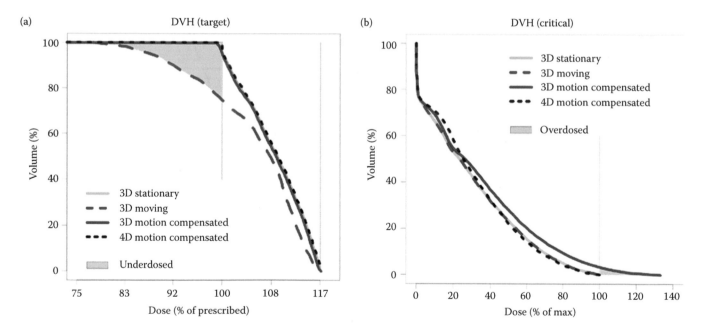

FIGURE 8.6 The dose distribution in target (a) and critical structure (b). Uncompensated target motion would lead to severe underdosing, and for motion compensation with 3D planning a very small deviation in the dose is cause by differences in the beam path between dose computation and moving beam. A more noticeable overdosing of the critical structure can be observed. Both deviations can be handled by 4D planning

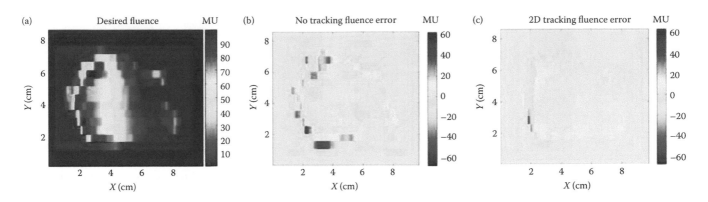

FIGURE 9.4 (a) Desired fluence distribution; (b) fluence error without motion compensation when the target experiences elliptical motion of 2.0 (parallel) × 0.5 cm (perpendicular); (c) fluence error with 3D synchronized motion tracking.

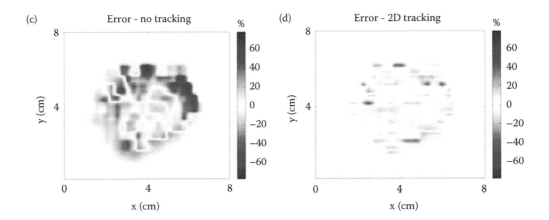

FIGURE 9.5 (a) Patient-derived 2D target motion in BEV; (b) Desired intensity map; (c) profile of error in intensity delivery without tracking; (d) profile of error in intensity delivery with 2D tracking in real time based on feedback; (e) dependence of delivery quality on the frequency of target position measurement; (f) dependence of delivery quality on real-time tracking latency of the system. See McMahon et al. (2008) for further details.

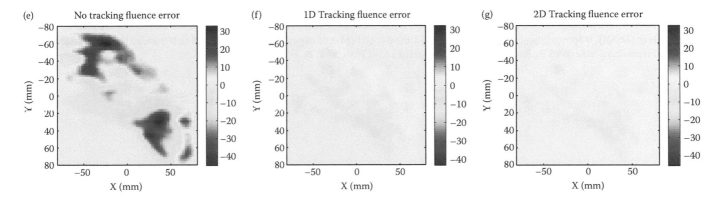

FIGURE 9.8 Optimal delivery parameters for a VMAT plan with 2D DMLC tracking as a function of time in a 360° single arc. (a) Gantry speed, (b) gantry acceleration, (c) MLC leaf velocity, and (d) dose rate. The discrepancies between the planned and delivered fluence for (e) no tracking, (f) 1D tracking, and (g) 2D tracking. See Rangaraj et al. (2008), Sun et al. (2010), and Appendix 9.B for details.

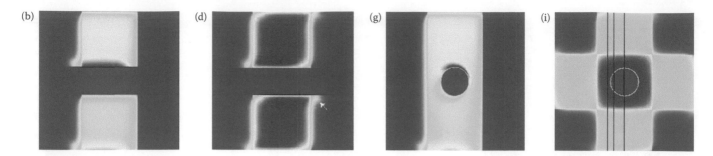

FIGURE 11.5 The impact of a 1.5 T transverse magnetic field on the dose distribution at a tissue–air interface. In (a), the setup of a layered water–air–water phantom is shown. In (b), the dose distribution from a single beam with the large dose increase at the interfaces is shown and in (c), the dose distribution from opposing beams is shown where the dose distribution in first order is homogeneous again. In (d), the setup for a cylinder of air in a homogeneous water phantom is shown. In (e), the ERE effect for this geometry is shown and (f) shows that 4 beams are required to restore the homogeneity for this geometry. (Adapted from Raaijmakers, A.J.E., Raaymakers, B.W., Lagendijk, J.J.W., *Phys. Med. Biol.*, 50:1363–1376, 2005.)

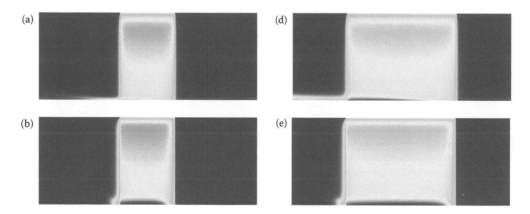

FIGURE 11.6 The impact of the magnetic field on the dose distribution is dependent on the magnetic field strength and the width of the radiation field. In (a) and (b), the dose distribution for a 5 × 5 cm² field is shown at 0.2 T and at 1.5 T. At 1.5 T, the ERE is much more pronounced than for 0.2 T although the same number of electrons are returning to the phantom. However, at 0.2 T, these electrons have large radii and enter the phantom again outside the radiation field. This is clearer in (c) and (d), where the dose distribution for a 10 × 10 cm² at 0.2 T and 1.5 T are shown. Now the amplitude of the ERE is approximately the same, but at 0.2 T the out-of-field dose deposition by the ERE is clearly seen on the left hand side of (c). (Adapted from Raaijmakers A.J.E., Raaymakers B.W., Lagendijk J.J.W., *Phys. Med. Biol.* 53(4): 909–23, 2008.)

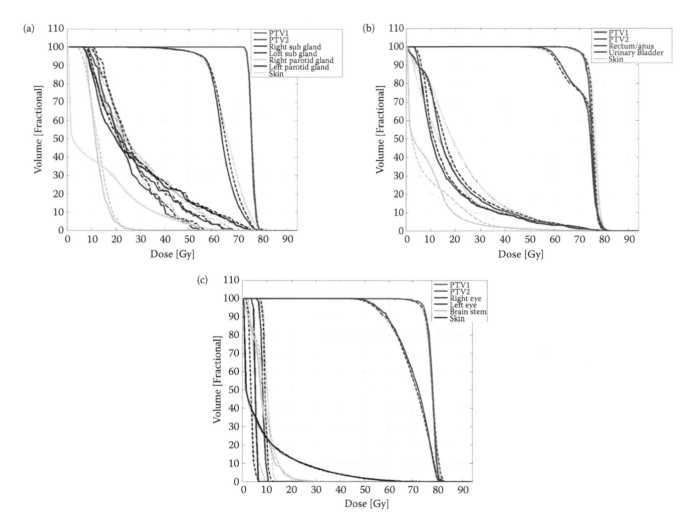

FIGURE 12.1 Cobalt treatment plans for 5 (dotted), 9 (dashed) and 71 (solid) equidistant coplanar treatment plans. The plans are for the (a) head and neck, (b) prostate, and (c) central nervous system.

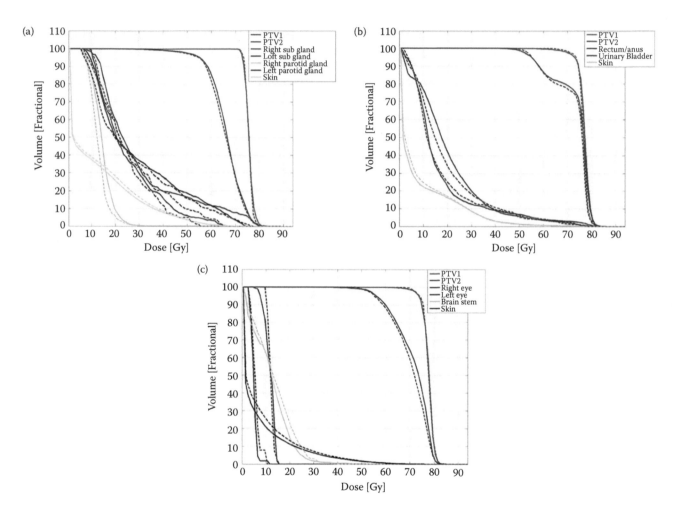

FIGURE 12.2 IMRT treatment plans for 6 MV (solid) and cobalt (dashed) 7 equidistant coplanar treatment plans. The plans are for the (a) head and neck, (b) prostate, and (c) central nervous system.

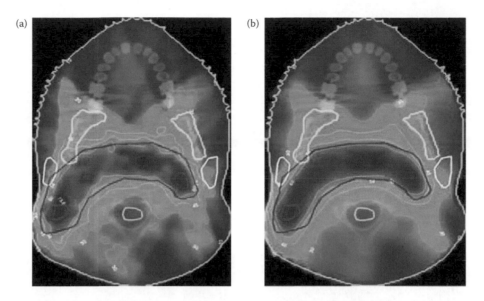

FIGURE 12.3 A Head and Neck treatment plan using 7 equally spaced fields for (a) 6 MV and (b) cobalt.

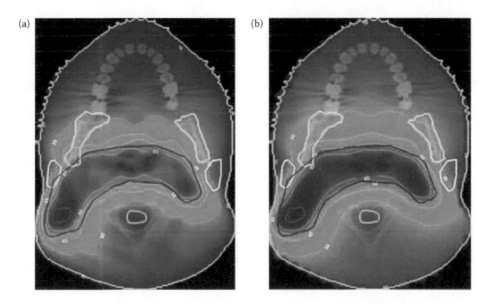

FIGURE 12.4 A Head and Neck treatment plan using 71 equally spaced fields for (a) 6 MV and (b) cobalt.

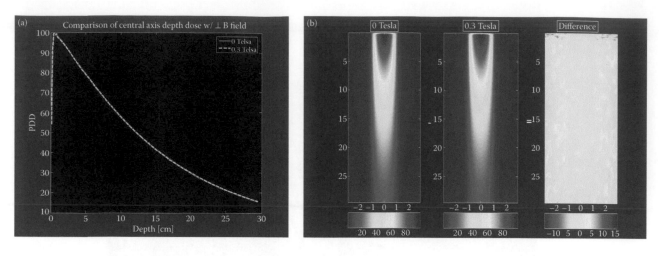

FIGURE 12.6 Comparison of (a) central axis depth dose, and (b) 2D distribution of a 2 cm circular beamlet striking a water phantom with and without a 0.3 T magnetic field applied perpendicular to the beam direction.

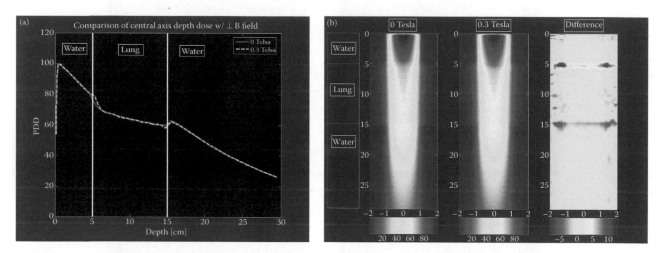

FIGURE 12.7 Comparison of (a) central axis depth dose, and (b) 2D distribution of a 2 cm circular beamlet striking a water/lung/water (5/10/15 cm, respectively) phantom with and without a 0.3 T magnetic field applied perpendicular to the beam direction.

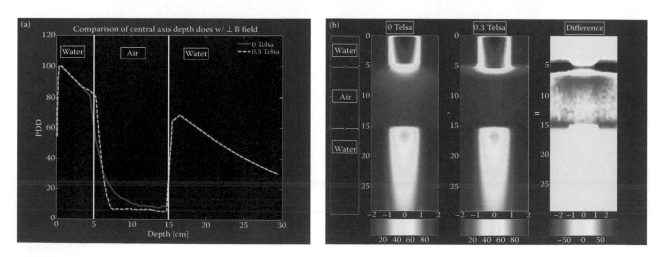

FIGURE 12.8 Comparison of (a) central axis depth dose, and (b) 2D distribution of a 2-cm circular beamlet striking a water/air/water (5/10/15 cm, respectively) phantom with and without a 0.3-T magnetic field applied perpendicular to the beam direction.

Robotic LINAC Tracking Based on Correlation and Prediction

7.1 Introduction 47
7.2 Correlation 48
    Basic Correlation Methods • Advanced Correlation Methods • Validation Experiment
7.3 Prediction of Respiratory and Pulsatory Motion 54
    The MULIN Family of Algorithms • The SVRpred Algorithm • Validation Experiments
7.4 Discussion 59
7.5 Outlook.. 60
    Fusion of Prediction and Correlation • Using Surrogates to Improve Prediction Quality
Additional Resources 61
References..61

Floris Ernst

Achim Schweikard

7.1 Introduction

In recent years, it has become increasingly practical to irradiate tumors while they move under the influence of respiration. In this process, the CyberKnife® system (Adler et al 1998) has won a firm spot in clinical treatment of moving tumors without using respiratory coaching, gating, or fixation. However, as has been discussed in Chapter 1, it is usually necessary to use surrogate signals to infer the tumor's changing position. To achieve this using the CyberKnife, the excursion of the patient's chest is recorded and correlated to artificial landmarks in the tumor observed during stereoscopic x-ray imaging (Schweikard et al 2000; Urschel et al 2007). (It has been demonstrated that this correlation process is actually feasible (Ahn et al 2004; Hoisak et al 2006).) This surrogate correlation model is subsequently used to guide the robotic arm carrying a linear accelerator to reduce or eliminate the effects of respiratory motion (see Chapter 4). Because the correlation between internal organ motion and the chest excursion may change, the model must be updated periodically.

There will be an inevitable systematic delay between the motion of the tumor and the motion of the treatment beam, which must be compensated. It has been shown previously that successful prediction of the time series stemming from human respiration is indeed possible using very different approaches. The algorithms investigated previously are based on Neural Networks, regression methods, and Kalman-filter approaches.

In this chapter, we will summarize methods currently in use or extensively researched for computing both the surrogate correlations and the temporal prediction of tumor motion during respiration. We will then present more advanced algorithms that

have the potential to improve the accuracy, reliability, and speed of correlation and prediction. For example, we propose to use support vector regression (SVR) (see Drucker et al 1997; Smola and Schölkopf 2004) to predict this time series. Additionally, to increase computational speed, the SVR function is iteratively built using the accurate online support vector (AOSVR) approach presented by Ma et al (2003). Much of our experimental evaluation of the algorithms has employed a porcine study.

Although we use the CyberKnife Synchrony platform as our exemplary clinical implementation, these algorithms are applicable to all real-time control systems used in respiratory motion management during radiation therapy.

Currently, the CyberKnife Synchrony correlation model is built using 10–20 measurements of internal fiducials acquired during the first couple of breathing cycles. The model is checked and updated periodically. This is typically done once every 2–5 minutes by taking another x-ray shot. Furthermore, external markers (currently three) are placed on the patient's chest at those points showing the greatest excursion. The currently employed correlation model is linear, curvilinear, dual-curvilinear, or a mixture of those and is based on the principal directional component of motion of each individual chest LED (Sayeh et al 2007).

We propose to improve the CyberKnife by modifying the correlation model employed in clinical practice. We compute the correlation using all three dimensions of movement of the LEDs as well as their first and second derivatives. This is done with a novel correlation model we have developed, which is based on e-support vector regression (e-SVR) (Drucker et al 1997). With this model, it is also possible to use multiple LEDs as input surrogates.

7.2 Correlation

The targeting method of choice implemented in the CyberKnife—fluoroscopic localization of the tumor—is not ideally suitable for continuous tracking. Due to the additional radiation dose delivered to the patient by imaging, an external surrogate signal is used to infer the target location based on sparsely acquired x-ray images. This process is called correlation and has been studied in great detail: the correlation between surrogate motion and target motion has been evaluated by Ahn et al (2004), Gierga et al (2005), and Koch et al (2004), linear correlation algorithms have been proposed and evaluated by Hoisak et al (2004), Hoisak et al (2006), Kanoulas et al (2007), Sayeh et al (2007), Schweikard et al (2000), Schweikard et al (2004), and Vedam et al (2003); more complex models based on Neural Networks (Murphy et al 2002; Yan et al 2006), SVR (Ernst et al 2009), Kalman filtering (Murphy et al 2002) and principal component analysis (PCA) (Khameme et al 2004) have also been examined. Another approach is to classify respiratory motion and fit a general model. This has been done by George et al (2005), in which sinusoidal models were evaluated. A more generic respiratory model using 4D-CT data has been created by McClelland et al (2006). Recently, a new look has been taken at monoscopic x-ray instead of the commonly used biplanar fluoroscopic approach (Cho et al 2008).

In this work, we focus on simple polynomial correlation as outlined by Sayeh et al (2007) and on SVR as proposed by Ernst et al (2009).

7.2.1 Basic Correlation Methods

All correlation methods* are based on one general assumption: there is a function f, which we can use to compute the position and orientation of the target region based on information acquired on the patient's chest or abdomen.

It is clear that, although some kind of function is bound to exist, the correlation between chest wall motion and interior organ motion cannot be expected to be completely deterministic. Even if we regard the body as a system of interacting elastic and inelastic tissues, we will not be able to theoretically model the correlation function for a single individual due to the complex nature of human motion.

The easiest approach to computing the correlation function f is to place radiopaque fiducials inside the target region and to attach markers to the patient's chest. Using a tracking system and two-plane fluoroscopy, the correlation between fiducial and marker motion can be determined. Subsequently, a polynomial model can be fitted to compute the position vector p_{Fi} of fiducial i from the position vector p_{Mj} of marker j, that is, $P_{Fi} = f(p_{Mj})$, where $f:\mathbb{R}^3 \to \mathbb{R}^3$ is a polynomial of arbitrary degree.

Since first experiments showed that motion on the chest is predominantly along one axis (perpendicular to the chest wall),

---

* The methods described in this section were presented by Sayeh et al. (2007) and are currently used clinically in Accuray's respiratory tracking system Synchrony™.

the initial simple correlation model was reduced to a polynomial $f:\mathbb{R} \to \mathbb{R}^3$. One problem of respiratory motion, however, is the fact that it exhibits hysteresis. This is due to differences in volume-to-pressure ratios during inspiration and expiration (West 2008). The polynomial model is able to capture this hysteresis if it is modified to use two polynomials, one for inspiration and one for expiration. We will call this model *dual polynomial*.

A severe limitation of the (dual) polynomial models, however, is the fact that they are built using sparse training data: in clinical settings, the patient is placed on the treatment couch, and five to twelve data points are used to create the model. It is clear that, using this few samples, it is nearly impossible to cover the patient's complete breathing cycle. Furthermore, since respiratory amplitudes change over time, it is reasonable to assume that measurements outside the data recorded for training will occur, resulting in possibly high errors, especially for higher order polynomials. Due to this reason, a fallback mechanism is implemented: in addition to the inspiration and expiration polynomials, a linear function is fitted to all training sets and it is used for correlation whenever measurements outside the range of the training data occur. Figure 7.1 shows an example of a dual quadratic correlation model built with (Figure 7.1b) and without (Figure 7.1a) fallback. The left panel also shows the linear correlation model.

7.2.2 Advanced Correlation Methods

Since the shortcomings of the simple correlation methods outlined in Section 7.2.1 are severe, new, and improved, correlation methods are required.

These correlation algorithms were developed and implemented in MATLAB®. The SVR machines were built using LibSVM (Chang and Lin 2001).

Let us assume that N is the number of samples we have taken. To build the correlation model, the input signal is divided into two parts: a training part $\mathcal{J} = \{1, \ldots, m\}$ and an evaluation part $\mathcal{E} = \{m+1, \ldots, N\}$. On the training part, we select points $\mathcal{M} \subseteq \mathcal{J}$ representative of the breathing pattern (i.e., points at maximum inspiration and expiration as well as points halfway between). Now, let $\mathbf{L}_{i,j,n}$ be the time series of the 19 IR LEDs ($i = 1, \ldots, 19$ is the number of the LED, $j = 1, \ldots, 3$ are the spatial coordinates and n is the temporal index), and let $\mathbf{F}_{k,j,n}$ be the time series of the four gold fiducials ($k = 1, \ldots, 4$ is the fiducial number, j and n as before).

The polynomial models are used to find coefficients of a linear or quadratic polynomial relating the principal directional component of motion of one LED to the principal directional component of motion of the fiducial.

A new correlation model can be based on ε-SVR. We do not only use the LEDs' principal directional component of motion as the polynomial models but all three dimensions. Second, information about the direction of breathing is directly built into the model by creating vectors $\mathbf{D}_i$ indicating the direction of breathing:

$$\mathbf{D}_{i,n} = \begin{cases} -1 & \text{for} \quad \tilde{\mathbf{L}}_{i,\cdot,n} - \tilde{\mathbf{L}}_{i,\cdot,n-1} < -0.05mm \\ 0 & \text{for} \quad -0.05mm \le \tilde{\mathbf{L}}_{i,\cdot,n} - \tilde{\mathbf{L}}_{i,\cdot,n-1} \le 0.05mm, n = 2, \ldots, N \\ 1 & \text{for} \quad \tilde{\mathbf{L}}_{i,\cdot,n} - \tilde{\mathbf{L}}_{i,\cdot,n-1} > 0.05mm \end{cases}$$

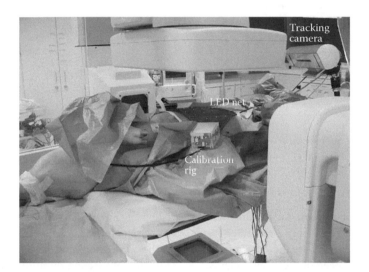

FIGURE 7.1 (a) linear (black) and dual quadratic (gray) models without fallback. (b) dual quadratic model (gray) with fallback to linear model. Both panels show the training data used (gray dots).

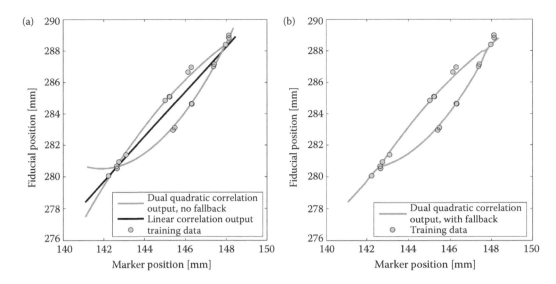

FIGURE 7.2 Experimental setup.

Here, $\widetilde{\mathbf{L}}_{i,\cdot,n}$ denotes the n-th sample of the principal directional component of the point cloud $\mathbf{L}_{i,\cdot,n}$. Third, we incorporate information about the signal's speed and acceleration by also bringing in the first and second derivatives $\mathbf{L}^{(1)}$ and $\mathbf{L}^{(2)}$ of the LEDs' positions. These derivatives are computed using central differences. Now, let

$$x_{i,m} = \left(\mathbf{L}_{i,m}^{T}, \mathbf{L}_{i,m}^{(1)\ T}, \mathbf{L}_{i,m}^{(2)\ T}, \mathbf{D}_{i,m}\right) \in \mathbb{R}^{9} \times \{-1,0,1\}$$

Then, for each $i = 1,\ldots,19$, $j = 1,\ldots,4$ and $m \in \mathcal{M}$ we create training samples $s_m^{i,j} = \{x_{i,m}, \widetilde{\mathbf{F}}_{j,m}\}$, i.e., the samples $s_m^{i,j}, m \in \mathcal{M}$, describe the relation between LED i and $\widetilde{\mathbf{F}}_j$, the principal directional component of motion of fiducial j. These samples are then used to train ε-SVR machines, which in turn serve as correlation models. The SVR machines are built using a Gaussian RBF kernel function.

7.2.3 Validation Experiment

To test the correlation algorithms, four gold fiducials were implanted into the liver of a living swine under ultrasound guidance. Figure 7.2 shows the experimental setup used. Respiratory motion of the liver was recorded in two sessions while the swine was ventilated manually using a bag valve mask. The swine was killed minutes prior to the experiments.

To acquire the fiducials' 3D position, a two-plane x-ray imaging device (Philips Allura Xper FD20/10) at the Institute for Neuroradiology (University Hospital Schleswig-Holstein, Lübeck) was connected to a high-resolution/high-speed frame grabbing system (Matrox Helios XA) to allow the capturing of live fluoroscopic video streams. The x-ray system takes images at a frame rate of 15 Hz.

To record the swine's chest surface motion, a net of 19 IR LEDs (see Knöpke and Ernst 2008 and Figure 7.3) was placed on

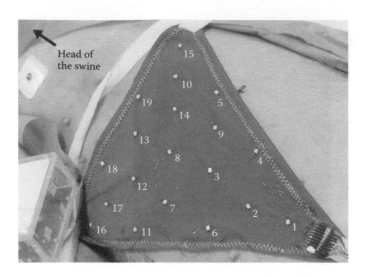

FIGURE 7.3   LED net.

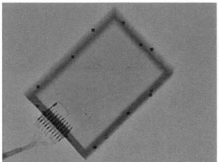

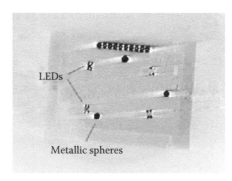

| (a) Frontal X-ray view | (b) Lateral X-ray view | (c) CT slice |

FIGURE 7.4   The calibration rig. (a) and (b) show frontal and lateral x-ray views, respectively, and (c) shows a CT slice of the rig.

the swine's abdomen. The LEDs were tracked using the Atracsys accuTrack 250* system, effectively delivering a recording frame rate of 216 Hz for each LED. The signals were then down sampled to 15 Hz to match the acquisition speed of the x-ray cameras.

To determine the geometric relation between the two x-ray imaging units and the tracking camera, a custom calibration rig was used: an acrylic box ($10 \times 7 \times 5$ cm³) with twelve embedded metallic spheres and eight LEDs was built. The system was calibrated by simultaneously acquiring an image of the calibration rig with the x-ray devices and determining the rig's position using the tracking camera. The actual calibration was performed using the POSIT algorithm (DeMenthon and Davis 1995), resulting in a projection error of less than one pixel (RMS). Both the LEDs and the metal spheres could be detected with submillimetre accuracy. Figure 7.4 shows frontal and lateral x-ray views as well as a CT slice of the calibration rig.

The frame grabber and the IR tracking system are connected to one machine (Intel Q9450, 8 GB RAM, CentOS 5 × 64).

To track the gold fiducials in the x-ray images, we developed a graphical tool kit written in C++ to perform

region-of-interest-based segmentation of ellipsoidal objects and triangulate the 3D position of the fiducials (Figure 7.5).

The correlation models were applied to two groups of signals (120 and 590 second duration) recorded during the ventilation of the swine. During both tests, not all LEDs were visible. In the first test, only LEDs 11, 14, and 16–19 were visible, in the second test, LEDs 6–14 and 16–19 were visible. The models were built using the first 20 seconds of motion. The ε-SVR's parameters were set to $C = 1$ and $\varepsilon = 0.2$ (signal 1) and $\varepsilon = 0.15$ (signal 2).

We also evaluated the influence of LED selection on correlation accuracy: first, which LED yields the best result and second, if the correlation model (Figure 7.6) can be improved by using more than one LED at the same time (Table 7.1).

7.2.3.1 Correlation Results

Analysis of LED motion shows that it is not only unidirectional but also does exhibit strong hysteresis relative to fiducial motion. This is the first indication that the simple polynomial models are not adequate.

The correlation plots of all the polynomial correlation models are given in Figure 7.7. Ideally, the curves would cover all dots. Clearly, the simple polynomial models do not fit the data very

---

* Atracsys LLC, Avenue du 24-Janvier 11, 1020 Renens, Switzerland.

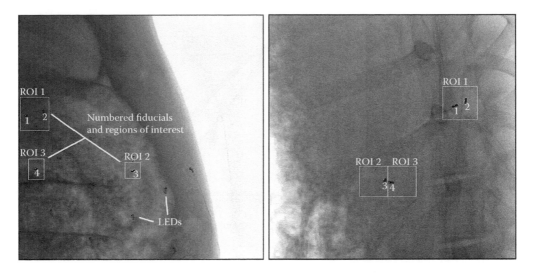

FIGURE 7.5 The tracking GUI. Both the fiducials and the LEDs are clearly visible. The regions used for segmentation are marked with white rectangles.

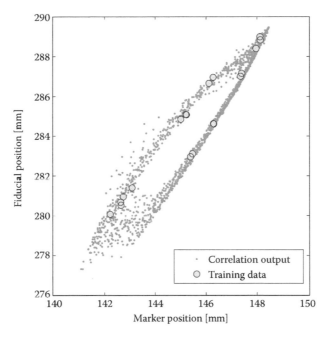

FIGURE 7.6 Output of the SVR correlation method.

TABLE 7.1 RMS Errors, 75% and 99% Confidence Intervals of the Correlation Models. Eleventh LED to First Fiducial

Model	RMS [mm]		75% CI [mm]		99% CI [mm]		Max [mm]	
	Sig. 1	Sig. 2	Sig. 1	Sig. 2	Sig. 1	Sig. 2	Sig. 1	Sig. 2
Linear	1.06	1.05	1.33	1.23	2.17	2.45	2.26	2.84
Quadratic	1.03	1.04	1.32	1.22	2.05	2.40	2.14	2.80
Quadratic, blended	1.03	1.04	1.32	1.22	2.05	2.40	2.14	2.80
Dual linear	0.78	0.96	0.91	1.16	1.78	2.25	4.36	5.88
Dual linear, blended	0.79	0.96	0.93	1.15	1.77	2.24	1.86	2.57
Dual quadratic	0.55	0.95	0.57	1.16	1.67	2.19	4.36	5.88
Dual quadratic, blended	0.50	0.95	0.52	1.14	1.48	2.18	1.71	2.54
ε-SVR	0.28	0.59	0.28	0.64	0.96	1.79	1.48	4.14

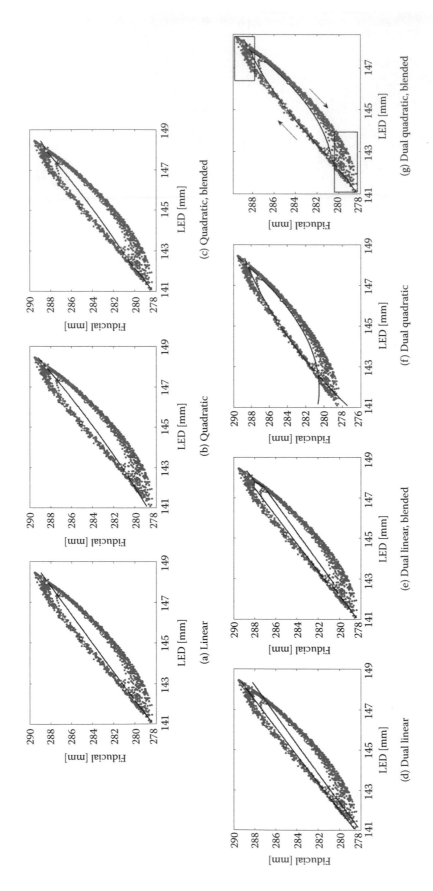

FIGURE 7.7 The polynomial models. The *x*-axis shows the principal directional component of the motion of LED 11, the *y*-axis shows the principal directional component of motion of the first fiducial. Model points are marked with circles, the model output is shown in black (a), (b), and (c) show the output of the simple models (no hysteresis). (d), (e), (f), and (g) show the dual models (with hysteresis). In (g), the boxes show the areas corresponding to maximum inspiration and expiration; the arrows indicate the breathing direction.

well, the dual polynomial models' matching is better. The actual numbers are given in Table 7.3.

Evaluation results for the best polynomial model (dual quadratic with blending) and for the ε-SVR model are given in Figure 7.9. The polynomial model does not only incur a larger RMS error (see Table 7.3) but also suffers from periodic errors at the inspiration and expiration peaks. The reason for this is that the model does not adequately capture correlation in the regions marked with black rectangles in Figure 7.7g.

Evaluation of the ε-SVR correlation model shows a much better matching: Figure 7.8 shows the correlation plots of the three axes of LED eleven versus the principal directional component of fiducial one. Clearly, the gray dots (output of the correlation model; Figure 7.9) correspond very well to the black dots (actual correlation). This is also reflected in the numbers given in Table 7.3: the SVR approach outperforms the best (dual) polynomial model by 45% (signal 1) or 38% (signal 2). Furthermore, the SVR model does not suffer from systematic errors like the polynomial models.

7.2.3.2 Selection of LEDs and Using Multiple LEDs

We also investigated the influence of LED selection on the quality of the correlation model. When selecting different LEDs as input surrogates, we see that on the first signal, the RMS error ranges from 0.27 to 0.38 mm; whereas, on the second signal, it ranges from 0.36 to 1.91 mm. Since the ε-SVR correlation model has been designed such that it can use input from more than one LED at a time, we evaluated the model for all possible pairs (triplets, quadruplets, etc). The results are shown in Figure 7.10.

We can see that on the first signal, using more than one LED does not noticeably improve correlation results. On the second signal, however, the best attainable correlation model uses six LEDs and outperforms the best model using one LED by 32%: the RMS value drops from 0.36 to 0.25 mm. In this case, the best polynomial model is outperformed by as much as 74%.

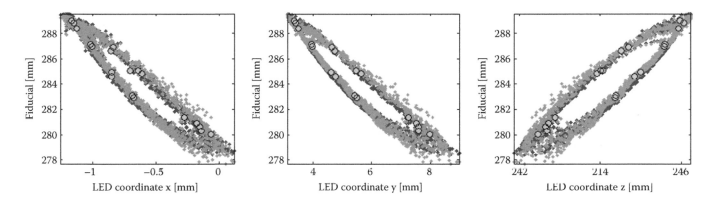

FIGURE 7.8 Output of the SVR correlation method (black dots: actual correlation; gray dots: model output).

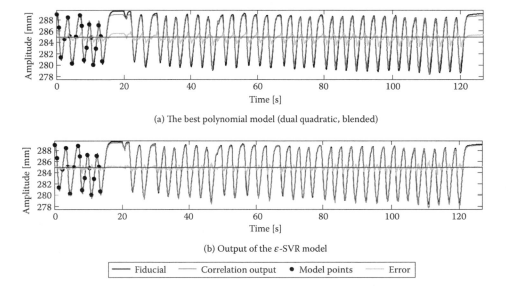

(a) The best polynomial model (dual quadratic, blended)

(b) Output of the ε-SVR model

— Fiducial   — Correlation output   • Model points   — Error

FIGURE 7.9 First test run, results of the correlation process. First 60 s are shown. Fiducial motion is shown in black, training points used as black circles and the correlation output in dark gray. The residual error is plotted in light gray. The respiratory pause around $t = 20$ s is accidental and not connected to the correlation model.

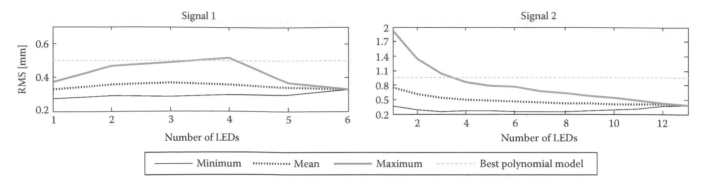

FIGURE 7.10 The plots show the range (minimum and maximum values) and mean of the RMS error when using multiple LEDs. Read: when using six LEDs (1716 possibilities) to build the correlation model on signal two, the resulting RMS error is, depending on the selected sextet, between 0.25 and 0.77 mm with a mean of 0.45 mm.

7.3 Prediction of Respiratory and Pulsatory Motion

Prediction of human respiratory motion has been investigated much since the beginning of robot-assisted surgery. The algorithms investigated so far can be classified according to their approach:

- Frequency tracking and modelling (Murphy et al 2002; Ramrath et al 2007)
- Kalman filtering (Murphy et al 2002; Sharp et al 2004; Sahih et al 2006; Ramrath et al 2007)
- Neural Networks (Murphy et al 2002; Sharp et al 2004; Murphy and Dieterich 2006; Sahih et al 2006; Murphy 2008; Goodband et al 2008; Murphy and Pokhrel 2009)
- Pattern Matching (Sayeh et al 2007)
- Fuzzy Prediction (Kakar et al 2005; Sheng et al 2007)
- Autoregressive Modelling (Murphy et al 2002; Sharp et al 2004; Vedam et al 2004; Sahih et al 2006; Ernst et al 2007; McCall and Jeraj 2007; Ren et al 2007; Sayeh et al 2007; Ernst and Schweikard 2008a; Ernst and Schweikard 2008b)
- SVR and Kernel Regression (Ruan et al 2007; Ernst and Schweikard 2009)
- Fusion or competition of multiple algorithms (Sheng et al 2007; Ernst et al 2008)
- Interacting multiple models (Putra et al 2008; Sahih et al 2006)
- Weighted frequency fourier linear combiner (Riviere et al 2001)

Only the algorithms described by Sayeh et al (2007) and Sheng et al (2007) are applied clinically. In this work, we will focus on two extremes: a very simple and fast algorithm, the MULIN family of algorithms (Ernst and Schweikard 2008b) and a very slow, complex algorithm, the SVRpred algorithm (Ernst and Schweikard 2009b).

7.3.1 The MULIN Family of Algorithms

The MULIN algorithms were introduced by Ernst and Schweikard (2008a and 2008b). The basic idea is to compute the predicted signal from an expansion of the error signal, that is,

the difference of the measured signal y and the signal $D(y,\delta)$, which is the signal y delayed by δ samples. The simplest member of this family of prediction algorithms, $MULIN_0$, is based on the following:

Assumption: Let y be the signal to predict and let $D(y,\delta)$ be the signal delayed by the prediction horizon δ. Then

$$y_{k+\delta} - y_k \approx y_k - y_{k-\delta} \tag{7.1}$$

Based on this, we define the simple linear prediction algorithm according to the following equation:

$$\hat{Y}^0_{k+\delta} = y_k - \Delta(y,\delta)_k \tag{7.2}$$

where $\Delta(y,\delta) = D(y,\delta) - y$.

Of course, this approach fails quite badly as soon as the above assumption does not hold. This, obviously, is the case whenever the signal's first derivative changes. The next step is therefore to further expand the prediction error $\Delta(y,\delta)_k$ to take this change into account. This is done by taking higher order differences, that is, by regarding the term $\Delta(y,\delta)_k$ in (7.2) as the unknown and to be predicted quantity. Applying (7.1) to $\Delta(y,\delta)_k$ results in the first-order prediction equation shown in

$$\hat{Y}^{1,M}_{k+\delta} = y_k - (\Delta(y,\delta)_k - \Delta(\Delta(y,\delta),M)_k)$$
$$= y_k - 2\Delta(y,\delta)_k + \Delta(y,\delta)_{k-M} \tag{7.3}$$

In this extended algorithm, a new parameter M was introduced. This parameter can be seen as something akin to the signal history length of LMS prediction algorithms: it controls how far back the algorithm should look to determine the change in the signal's first derivative. Naturally, we would expect M to be equal to δ since we are trying to predict the change of the difference signal $\Delta(y,\delta)$. In those cases, however, where the signal we try to predict is highly irregular or instable either due to the presence of noise or high signal variation or where the sampling rate is low, it is reasonable to assume that the correlation between y_{k-M} and $y_{k+\delta}$ is higher as the M becomes smaller. This is highlighted in Figure 7.11.

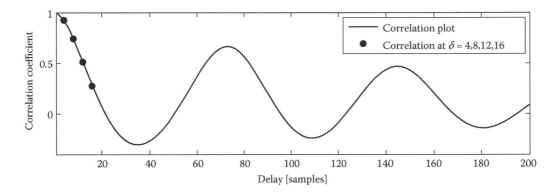

FIGURE 7.11 Autocorrelation of a real breathing motion signal.

Nevertheless, when both sampling rate and signal-to-noise ratio are high, we can assume that long-term signal dependencies exist. To exploit these dependencies, we repeat the expansion process, generating second and third order linear prediction algorithms. This is done by applying (7.2) to $\Delta(y,\delta)_k$, $\Delta(y,\delta)_{k-M}$ and $\Delta(y,\delta)_{k-2M}$.

$$\hat{Y}_{k+8}^{2,M} = y_k - 4\Delta(y,\delta)_k + 4\Delta(y,\delta)_{k-M} - \Delta(y,\delta)_{k-2M}$$
$$\hat{Y}_{k+8}^{3,M} = y_k - 8\Delta(y,\delta)_k + 12\Delta(y,\delta)_{k-M} - 6\Delta(y,\delta)_{k-2M} + \Delta(y,\delta)_{k-3M}$$
$$(7.4)$$

Clearly, in those cases where the correlation between the quantities $y_{k+\delta}$ and $(y,\delta)_{k-2M}$ as well as between $y_{k+\delta}$ and $\Delta(y,\delta)_{k-3M}$ is low, we do not expect improvement over the simple or first-order prediction algorithms, and, in fact, might even witness degradation of the prediction performance.

Theorem. The preliminary MULIN algorithms given in Equations (7.2) to can be combined into

$$\hat{Y}_{k+\delta}^{n,M} = y_k + \sum_{i=0}^{n} \frac{(-1)^{i+1} 2^{n-i} n!}{i!(n-i)!} \Delta(y,\delta)_{k-iM} \qquad (7.5)$$

Furthermore, by repeating the expansion process, this formula also holds for higher order algorithms.

Proof. The theorem is proved by induction on n. For $n = 0$, we get

$$\hat{Y}_{k+\delta}^{0,M} = y_k + \frac{(-1)^{(0+1)} 2^{(0-0)} 0!}{0!(0-0)!} \Delta(y,\delta)_{k-0\cdot M} = y_k - \Delta(y,\delta)_k$$

which is the formula given in (7.2). Now, assume the formula to be correct up to n. We need to show that applying the expansion rule given in (7.2) to (7.5) gives the same result as replacing n by $n+1$ in; that is,

$$\sum_{i=0}^{n} \frac{(-1)^{i+1} 2^{n-i} n!}{i!(n-i)!} (2\Delta(y,\delta)_{k-iM} - \Delta(y,\delta)_{k-(i+1)M})$$
$$= \sum_{i=0}^{n+1} \frac{(-1)^{i+1} 2^{n+1-i} (n+1)!}{i!(n+1-i)!} \Delta(y,\delta)_{k-iM}.$$

This is done by comparing coefficients of $\Delta(y,\delta)_{k-iM}$:

$$\Delta(y,\delta)_k$$

$$\frac{(-1)^{(0+1)} 2^{(n-0)} n!}{0!(n-0)!} \cdot 2 = -2^{n+1}; \quad \frac{(-1)^{(0+1)} 2^{(n+1-0)} (n+1)!}{0!(n+1-0)!} = -2^{n+1}.$$

$$\Delta(y,\delta)_{k-iM}, i=1,\dots,n$$

Left:

$$\frac{(-1)^{i-1+1} 2^{n-i+1} n!}{(i-1)!(n-i+1)!} \cdot (-1) + \frac{(-1)^{i+1} 2^{n-i} n!}{i!(n-i)!} \cdot 2 = \frac{(-1)^{(i+1)} 2^{(n+1-i)} n!}{(i-1)!(n+1-i)!}$$
$$+ \frac{(-1)^{i+1} 2^{n+1} - i n!}{i!(n-i)!} = \frac{i(-1)^{i+1} 2^{n+1-i} n! + (n+1-i)(-1)^{i+1} 2^{n+1-i} n!}{i!(n+1-i)!}$$
$$= \frac{(-1)^{i+1} 2^{n+1-i} (n+1)}{i!(n+1-i)} = \frac{(-1)^{i+1} 2^{n+1-i} (n+1)!}{i!(n+1-i)!}$$

Right:

$$\frac{(-1)^{i+1} 2^{n+1-i} (n+1)!}{i!(n+1-i)!}$$

$$\Delta(y,\delta)_{k-(n+1)M}$$

Left:

$$\frac{(-1)^{n+1} 2^{(n-n)} n!}{n!(n-n)!} (-1) = (-1)^{n+2}$$

Right:

$$\frac{(-1)^{n+1+1} 2^{n+1-n-1} (n+1)!}{(n+1)!(n+1-n-1)!} = (-1)^{n+2}$$

This completes the proof.

To exploit long-term dependencies, an exponential smoothing parameter $\mu \geq 0$ is introduced to form the final form of the MULIN algorithms:

$$\hat{y}_{n+\delta}^{\mathrm{MULIN}_{k,M}} = \mu \cdot \hat{Y}_{n+\delta}^{k,M} + (1-\mu) \cdot \hat{y}_{n+\delta-1}^{\mathrm{MULIN}_{k,M}}, n = 0,\dots,4$$
$$(7.6)$$

Here, M is the algorithmic parameter controlling the memory step size of the MULIN algorithms.

7.3.2 The SVRpred Algorithm

Recently, SVR has been used to predict respiratory motion (Ernst and Schweikard 2009a,b). The idea behind this, given some kind of training data $\{(u_{k_1}, y_{k_1+\delta}), \ldots, (u_{k_L}, y_{k_L+\delta})\}$, is to find a function $f(x)$ that has at most ε deviation from the actually obtained targets $y_{i+\delta}$ for all the training data.

Here, u_i is some signal history vector from $\mathbb{R}^M$. The components of u_i are selected in some manner from the past part of the time series y, i.e. $u_{i,j} = y_{g(i,j)}$ where g is strictly monotonously decreasing in i, $g(i, 1) = i$ and $g(i, j) \geq 1$ for all j. In general, the function f can be written as

$$f(x) = w^{\mathrm{T}} \Phi(x) + b \tag{7.7}$$

for a possibly nonlinear function Φ mapping $\mathbb{R}^M$ to a feature space $\mathcal{F}$, a vector $w \in \mathcal{F}$ and a scalar $b \in \mathbb{R}$. This function is now computed by solving the optimization problem:

$$\min_{w,b} \frac{1}{2}\|w\|^2 + C \sum_{i=1}^{L} \xi_i + \xi_i^*$$

such that

$$
\begin{aligned}
y_{i+\delta} - w^{\mathrm{T}}\Phi(u_i) - b &\leq \varepsilon + \xi_i \\
w^{\mathrm{T}}\Phi(u_i) + b - y_{i+\delta} &\leq \varepsilon + \xi_i^* \\
\xi_i, \xi_i^* &\geq 0, i = 1, \ldots, L.
\end{aligned}
\tag{7.8}
$$

Here, ξ_i and ξ_i^* are slack variables introduced to cope with training data possibly violating the condition $|f(u_i) - y_{i+\delta}| \leq \varepsilon$. C controls how much these deviations are penalized. Following Smola and Schölkopf (2004), we can formulate the Lagrangian function using Lagrange multipliers $\alpha, \alpha^*, \eta,$ and η^*:

$$
\begin{aligned}
\mathcal{L} = &\frac{1}{2}\|w\|^2 + C\sum_{i=1}^{L}(\xi_i + \xi_i^*) - \sum_{i=1}^{L}(\eta_i \xi_i + \eta_i^* \xi_i^*) \\
&-\sum_{i=1}^{L}\alpha_i(\varepsilon + \xi_i - y_{i+\delta} + w^{\mathrm{T}}\Phi(u_i) + b) \\
&-\sum_{i=1}^{L}\alpha_i^*(\varepsilon + \xi_i^* + y_{i+\delta} - w^{\mathrm{T}}(u_i) - b)
\end{aligned}
\tag{7.9}
$$

From the saddle point condition follows that the partial derivatives of $\mathcal{L}$ with respect to the primal variables b, w, ξ and ξ^* have to vanish at optimality.

By substituting this into (7.9) and taking the nonnegativity of the dual variables into account, we subsequently arrive at the dual optimization problem

$$
\begin{aligned}
\max_{\alpha,\alpha*} &\frac{1}{2}\sum_{i,j=1}^{L}\Phi(u_i)^{\mathrm{T}}\Phi(u_j)(\alpha_i - \alpha_i^*)(\alpha_j - \alpha_j^*) - \varepsilon\sum_{i=1}^{L}(\alpha_i + \alpha_i^*) \\
&+\sum_{i=1}^{L}y_{(i+\delta)}(\alpha_i - \alpha_i^*) \\
&\text{s.t. } 0 \leq \alpha_i, \alpha_i^* \leq C, \sum_{i=1}^{L}(\alpha_i - \alpha_i^*) = 0, i = 1, \ldots, L.
\end{aligned}
\tag{7.10}
$$

In accordance with Smola and Schölkopf (2004), we introduce a kernel function $k(x,y)$ as

$$k(x,y) = \Phi(x)^{\mathrm{T}}\Phi(y)$$

In the end, the creation of an SVR function f boils down to selecting a kernel function and solving the optimization problem given in (7.10).

7.3.2.1 Prediction of Human Respiratory Motion

Up to now, prediction of human respiratory motion using kernel-based machine learning methods has not been investigated much. One reason is that, until very recently, prediction with SVR required either constant retraining of the SVR function or the assumption of stationarity. With Ma et al (2003), however, it has become possible to implement SVR for online prediction by iteratively forgetting older samples and adding new measurements to the SVR function without having to completely retrain it. Still, problems remain:

- Choice of kernel function and corresponding parameters.
- Selection of the SVR parameters C, the penalty factor, and ε, the error insensitivity level.
- Selection of the signal history to avoid the curse of dimensionality (Verleysen and Francois 2005).
- Computational time is still very high and might prohibit real-time applications.

The proposed prediction algorithm, called SVRpred, was implemented in C++ using the AOSVR library created by Francesco Parella (Parrella 2007). Available kernel functions are

- Linear, i.e. $k(x,y) = \langle x,y \rangle$
- Polynomial, i.e. $k(x,y) = (\langle x,y \rangle + 1)^\sigma$,
- Gaussian RBF, i.e. $k(x,y) = \exp\left(-\dfrac{\|x-y\|_2^2}{2\sigma^2}\right)$
- Exponential RBF, i.e. $k(x,y) = \exp\left(-\dfrac{\|x-y\|_1^2}{2\sigma^2}\right)$
- Multilayered perceptron, i.e. $k(x,y) = \tanh(\sigma\langle x,y \rangle + \tau)$

Here, the values σ and τ are called kernel parameters. These parameters can be selected arbitrarily for all kernels but the MLP kernel. Conditions for the MLP kernel's parameters are given by Burges (1999).

7.3.2.2 The AOSVR Algorithm

The AOSVR library implements the algorithm for incremental and decremental support vector learning introduced in Ma et al (2003). From (7.7) and (7.10) follows that the ε-insensitive loss function $f(x)$ can be determined as

$$f(x) = \sum_{i=1}^{L}(\alpha_i - \alpha_i^*)k(u_i, x) + b$$

where L is the number of samples trained, α_i and α_i^* are the Lagrangian multipliers associated with the sample pair $(u_i, y_{i+\delta})$, and k is the kernel function used. Let $\theta_i = \alpha_i - \alpha_i^*$. Note that, due to the Karush–Kuhn–Tucker (KKT) conditions derived from

(7.10), at most one of α_i and α_i^* can be nonzero and both must be nonnegative. Hence they both are determined uniquely by the value of θ_i. Adding a new sample $(u_c, y_{c+\delta})$ to an already trained SVR function then works as follows:

- Start off with $\theta_c = 0$. The new sample does not yet contribute to the cost function.
- Iteratively change θ_c until the new sample meets the KKT conditions corresponding to the dual SVR optimization problem.
- While doing so, ensure that the already trained samples continue to satisfy the KKT conditions. This is achieved by changing the corresponding θ's and – if necessary – by moving samples from the support to the noncontributing set (or vice versa).

Conversely, removing an already trained sample $(u_c, y_{c+\delta})$ from the SVR function is done in a similar fashion: the corresponding coefficient θ_c is iteratively reduced to zero—it then does not contribute to the cost function f—while maintaining KKT conditions for all samples in the SVR function. Due to the complexity of the incremental and decremental algorithms, the reader is referred to the original work of Ma et al. (Ma et al 2003) for further explanation.

7.3.2.3 Selecting the Signal History

Since the proposed prediction method is based on regression, we need to select a subset of previous samples to use as input vector for the SVM. Naturally, one would start with selecting a signal history length M, ideally a multiple of the prediction horizon δ, and take all old samples between t_n, the current position in time, and t_{n-M+1} to try predict $t_{n+\delta}$ from these. In our application, however, when we deal with high-frequency sampling, we may have a prediction horizon of up to 200 samples (see Knöpke and Ernst 2008) and thus end up trying to model an extremely high dimensional signal even when using moderate signal history lengths of, for example, 2δ.

Furthermore, it seems reasonable to assume that the intrinsic dimension of the regressor of dimension M is probably less than M. Unfortunately, we cannot employ the tools proposed by Verleysen and Francois (2005) to reduce the dimensionality of the regressor (such as PCA, Curvilinear Component Analysis (Demartines and Herault 1997), Curvilinear Distance Analysis (Lee et al 2002), and Isomaps (Tenenbaum et al 2000) since the sensory input data available changes as time progresses and we might thus face a change of intrinsic dimension when breathing pattern changes. We therefore decided to take a more conventional approach: to reduce the dimension of the regressor by simply skipping input samples. Therefore, the signal history is either linearly or quadratically spread out; that is, we do not take the samples at $t_n, t_{n-1}, t_{n-2}, \ldots, t_{n-M+1}$, but we take either the samples at $t_n, t_{n-l}, t_{n-2l}, \ldots, t_{n-(M-1)l}$ or at $t_n, t_{n-l}, t_n-(2_l)^2, t_n-(3_l)^2, \ldots, t_{n-((M-1)l)}^2$, where l is called the stepping parameter.

7.3.2.4 Prediction Method

Using these simplifications, the initial version of the proposed prediction method is outlined in Listing 1.

```
FOR n = δ; n <NumberOfSamples; n++
  ' build signal history
  ' mode can be LINEAR or QUADRATIC
  u_n = HISTORY(n,M,mode,stepping)
  ' train
  TRAIN(u_{n-δ},y_n)
  ' predict
  ŷ_{n+δ} = PREDICT(u_n)
  ' forget
  IF n > MinimalNumberOfTrainedSamples
    FORGET(sample(0))
  END
END
```

LISTING 1   Initial version of the SVRpred prediction algorithm

7.3.2.5 Optimization for Speed

One drawback of the SVRpred method is that on today's business PCs, computational speed is not high enough. To evaluate and possibly alleviate this problem, we propose to train and forget only a subset of all samples instead of training each new measurement and subsequently forgetting the oldest sample. This is done according to Listing 2.

```
τ = 0.1 ' training percentage
FOR n = δ; n <NumberOfSamples; n++
  ...
  ' training
  IF n ≡ 0(mod INT(1/ τ))
    TRAIN(u_{n-δ},y_n)
  END
  ' forgetting
  IF n > MinimalNumberOfTrainedSamples
    IF n ≡ 0(mod INT(1/ τ))
      FORGET(sample(0))
    END
  END
  ...
END
```

LISTING 2   Training and forgetting only a subset of samples

7.3.3 Validation Experiments

The SVRpred and MULIN algorithm were tested using several types of signals, using a prediction horizon of 150 ms. To achieve this, they were implemented in C++, using a general prediction framework (Rzezovski and Ernst 2008). For timing purposes, the evaluations were run on a machine running CentOS 5 x64, using one core of an Intel Q9450 processor. Available RAM was 8 GB. First, we used the simulated breathing signal described in (7.11). This signal thus has a sampling rate of 100 Hz and the required prediction horizon of 150 ms corresponds to 15 steps in time.

$$y = 2\sin(0.25\pi \cdot t)^4, \, t = (0, 0.01, \ldots, 100)^{\mathrm{T}} \qquad (7.11)$$

Second, the simulated signal was corrupted with Gaussian noise of zero mean and a standard deviation of $\sigma = 0.025$

In a third experiment, the algorithm was tested on the real breathing signal shown in Figure 7.12. This signal is equidistantly sampled at a rate of 26 Hz and has a length of 7,500 sampling points (just less than five minutes). The required prediction horizon thus corresponds to four steps in time.

In all cases, the relative RMS error (7.12) was computed and used as a measure for the quality of the prediction.

$$\text{RMS}_{\text{rel}} = \frac{\left\| y - \hat{y} \right\|}{\left\| y - D(y, \delta) \right\|} \qquad (7.12)$$

7.3.3.1 Parameter Selection

As mentioned before, the selection of the parameters of the SVRpred algorithm may be very difficult. Especially selecting the training insensitivity level ε and the signal history length M (and the stepping parameter l) are not straightforward. Since the optimal values for these parameters—as well as for C, σ, and τ—cannot be determined in the setting of online prediction, we selected the values as follows:

- C was set to the default value of 30.
- σ was set to the default value of 2.
- ε was set to the standard deviation of the noise present in the signal. On the simulated signal, this value is known and ε was thus set to 0.025. On the real signal, however, this value is unknown. We resorted to approximating this value according to our previous work (see Ernst et al 2010a).

$$\sigma \approx \text{std}((W_{1,1}, \ldots, W_{1,N})^{\text{T}})$$

Here, W_1 is the first scale of the à trous wavelet decomposition of the signal y. We thus set ε to 0.0355 on the real signal.

- M was selected by looking at the correlation coefficients of the signal and its lags. Let $x_m = \Delta(y, m + \delta)$. We then computed

$$\chi_m = \frac{\sum_{i=1}^{N-\delta-m}(x_{m,i} - \bar{x}_m)(y_i - \bar{y})}{\sqrt{\sum_{i=1}^{N-\delta-m}(x_{m,i} - \bar{x}_m)^2 \sum_{i=1}^{N-\delta-m}(y - \bar{y})^2}}$$

i.e., the autocorrelation coefficient for different lags $m = 1, \ldots, 200$. Here, $\bar{x}_m$ and $\bar{y}$ denote the mean values of x_m and y, respectively. We then selected M such that $\chi_m \geq 0.6$ for all $m = 1, \ldots, M$. This resulted in $M = 5$ on the real and $M = 39$ on the simulated signal.

- l was set to 1, i.e., we used all points of the signal history in building the SVR function.
- τ was set to 0.1.

To determine M and ε, only the first 2,000 samples of the two signals were used. In turn, these 2,000 samples were disregarded when computing the prediction performance. All tests were performed using the Gaussian RBF kernel function.

7.3.3.2 Evaluation Results

Using the parameters selected as outlined above, we computed the predicted signals using the SVRpred algorithm. The results were compared to the optimal results of the MULIN algorithms, that is, their parameters were selected using a global optimization on the complete signals. Of course, this optimization cannot be done in reality, but for comparison purposes, it is valid. Table 7.2 shows the prediction results.

As outlined in Section 7.3.1.6, the selection of the SVRpred algorithm's parameters was somewhat arbitrary. We have also evaluated the performance of the algorithm for changes in C

TABLE 7.2   Prediction Results

Predictor	Relative RMS	Parameters	Evaluation Time
Simulated Signal, No Noise			
MULIN	0.02	$k = 2, \mu = 0.70, M = 5$	0.01
SVRpred	0.02	$\varepsilon = 0.001^{\text{a}}, M = 39, l = 1, \tau = 0.1$	92.93 [b]
Simulated Signal, Added Noise			
MULIN	0.27	$k = 1, \mu = 0.228, M = 21$	0.01
SVRpred	0.18	$\varepsilon = 0.025, M = 39, l = 1, \tau = 0.1$	3.07
Real Signal			
MULIN	0.64	$k = 2, \mu = 0.52, M = 1$	0.02
SVRpred	0.54	$\varepsilon = 0.0355, M = 5, l = 1, \tau = 0.1$	1.60

[a] Note that ε must be larger than 0
[b] The AOSVR algorithm had problems with incremental and decremental learning

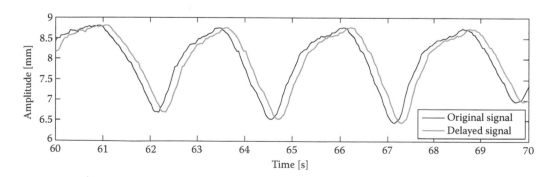

FIGURE 7.12   Breathing signal of a CyberKnife patient and the signal delayed by 150 ms (sampled at 26 Hz).

and σ, where we found that changes in *C* do not result in strong changes of the RMS error, while changes in σ do have some influence.

Second, we have also evaluated the effect of changes in *M* and (with constant *C* and σ) on both prediction performance and evaluation time. We found that the optimal values for *M* and ε are relatively close to the values selected before: ε changes from 0.0355 to 0.04, *M* changes from 5 to 6, and the resulting relative RMS hardly changes at all: from 0.5419 to 0.5387. Similar results hold for the simulated signal with noise: here, the optimal values are ε = 0.02 and *M* = 39 with *l* = 10. The relative RMS drops from 0.1897 to 0.1502.

The predicted signals of the simulated signal with noise and the real signal, using optimal parameters, are shown in Figure 7.13. Note that, especially on the real signal, the algorithms perform similarly during inspiration and expiration whereas the SVRpred algorithm clearly outperforms the MULIN algorithms at the inhalation peaks.

7.3.3.3 Influence of the Training Percentage

For the sake of clarity, in the above section, all evaluations were done using a training percentage of τ = 0.1, that is, using every sample u_n such that $n \equiv 0 \bmod(1/\tau)$. Of course, the speed of the prediction can be increased by lowering this percentage and, presumably, prediction accuracy will increase in turn.

This is confirmed by evaluating the prediction results on the real signal for τ in the range from 0.001 to 0.25 as shown in Figure 7.14.

We can see that the best performance can be obtained using a training percentage of τ = 0.25, resulting in a relative RMS of 0.49, requiring 78.77 seconds of evaluation time. This corresponds to approximately 10 ms of computation time per sample. Since the sampling rate of the real signal is 26 Hz (i.e., 38.5 ms spacing), the algorithm can still be used in an online prediction setting.

7.3.3.4 Batch Training and Prediction

From Ma et al (2003), we know that the AOSVR training algorithm produces the exact same SVR function as batch training does. Therefore, the repeated application of a batch training method (like Chang and Lin 2001) will not yield different prediction results. Ma et al (2003) compared the prediction speed of the AOSVR algorithm to the LibSVM algorithm. The result was that, in every case, the online approach was faster. We thus did not perform any tests using LibSVM. One question remains, however: how does the SVRpred algorithm perform when no additional training (i.e., learning of new samples or forgetting of old samples) is done after the initial training epoch of 2,000 samples. Results are clear: on all test signals, the relative RMS error of the static predictor is worse than the relative RMS of the updated predictor. Details are given in Table 7.3.

7.4 Discussion

It has been shown that the new prediction algorithm presented, SVRpred, is capable of delivering better prediction performance

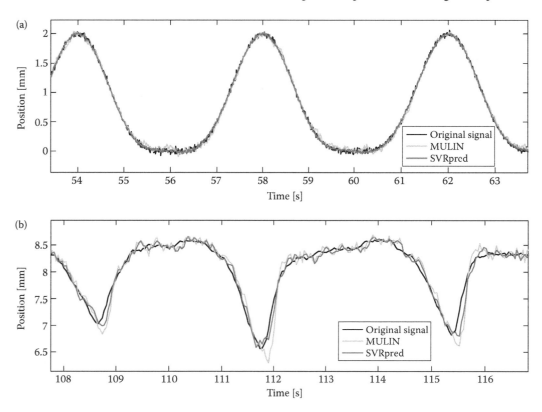

FIGURE 7.13 **(See color insert.)** Results of the MULIN and SVRpred algorithms on the simulated signal with noise (a) and the real signal (b). The original signal is shown in black, the MULIN algorithm in blue and the SVRpred algorithm in red.

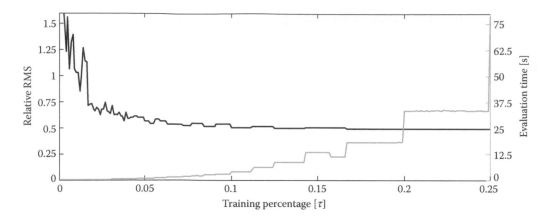

FIGURE 7.14 Evaluation of the impact of training percentage on the relative RMS (black) and evaluation time (gray), real signal used. Other parameters are the optimal values shown in Table 7.2

TABLE 7.3 Comparison of Static and Updated SVRpred on the Three Test Signals

Signal	Static Time	Rel. RMS	Updated Time	Rel. RMS
Simulated, no noise	4.07	0.02	92.93	0.02
Simulated, added noise	0.56	0.19	3.07	0.18
Real	0.27	0.77	1.60	0.54

than previous algorithms. The improvement is considerable: on simulated data, the MULIN algorithms are outperformed by up to 13% points, and on real data they are outperformed by up to 16% points. The main drawback of the algorithm is its slowness: even with the proposed optimization for speed, predicting a signal may take very long. Therefore the algorithm can—nowadays—only be employed for live prediction of signals sampled at a low rate.

While the improvement of the RMS error may seem small, it is of clinical relevance: the output of the prediction algorithm is used in a correlation model to determine the position of the tumor. It has been shown that the internal organ motion is potentially much larger than the motion of the skin (Ahn et al 2004). Hence the errors from previous processing steps will be amplified considerably by the correlation model.

The fact that correlation on the second signal (with a length of 590 seconds) can be improved significantly by using multiple LEDs is due to the signal's complexity: ventilation speed was varied as was the ventilation volume such as to mimic changing breathing patterns. It seems that especially the changes in breathing mode cannot be adequately covered by using one single LED.

We also went beyond the work presented by Tang et al (2004) and Seppenwoolde et al (2007). Tang et al (2004) proposed that the internal position of the tumor was measured with NDI's magnetic tracking system using biopsy needles inserted into the swine. This approach suffers from poor temporal and spatial resolution and also from possibly altering the organ's motion patterns due to the insertion of biopsy needles. According to

Seppenwoolde et al (2007), the evaluation of the correlation was done retrospectively on lung tumor patients, in which the tumor was located using stereoscopic fluoroscopy at changing angles and the external motion was recorded using a laser-based measuring device. The disadvantages in this study are the relatively short duration of the recordings (just over 1 minute in average) and the fact that the laser-based measuring system can only record distances and does not show the 3D displacement of a single point in space.

In the near future, we plan to continue this study with more swines either under mechanical lung ventilation or under free breathing. We also hope to see this new correlation model implemented in the CyberKnife to perform tests under real conditions. Also, further investigation needs to be done as to where to place the LEDs on which the model is built, since this placement significantly influences the correlation accuracy.

7.5 Outlook

In the future, we will look into the possibility of using feature maps or other dimensionality reduction methods to improve the algorithm's speed and accuracy. In this context, we will also further investigate the difficulties of proper parameter selection.

Other possible research areas are fusion of prediction and correlation and the inclusion of surrogates to improve prediction quality.

7.5.1 Fusion of Prediction and Correlation

In this area, some work has been done in the past: an autoregressive and a Neural Networks approach for combined prediction and correlation has been evaluated by Murphy et al (2002), and another combined approach using Neural Networks has been proposed by Yan et al (2006). These methods, however, have not been pursued much in the last years.

We believe that a combination of the SVRpred prediction and correlation algorithms outlined in this work is promising. This will be investigated in the future.

7.5.2 Using Surrogates to Improve Prediction Quality

Another novel approach to improve prediction quality is to use other surrogate signals. In the context of respiration, it might be beneficial to include data from aeroplethysmography. Since these surrogates can be incorporated easily into the SVRpred algorithm, evaluating this idea is not very complex.

In the context of cardiac motion prediction, which is required for—for example—beating heart cardiac surgery, we have successfully used the ECG as an input surrogate for the SVRpred algorithm. Details can be found in Ernst et al 2010b.

Additional Resources

The data used for the experiments, as well as the prediction toolkit, are available for public download from http://signals.rob.uni-luebeck.de.

References

J. R. Adler Jr., A. Schweikard, M. J. Murphy, and S. L. Hancock. Image-guided stereotactic radiosurgery: The CyberKnife, pp. 193–204 in G. H. Barnett, D. Roberts, and B. L. Guthrie, editors, *Image-Guided Neurosurgery: Clinical Applications of Surgical Navigation*. Quality Medical Publishing, St. Louis, MO 1998.

S. Ahn, B. Yi, Y. Suh, J. Kim, S. Lee, S. Shin, and E. Choi. A feasibility study on the prediction of tumour location in the lung from skin motion. *The British Journal of Radiology*, 77:588–596, 2004. DOI: 10.1259/bjr/64800801.

C. J. C. Burges. Geometry and invariance in kernel based methods, pp. 89–116 in B. Schölkopf, C. J. C. Burges, and A. J. Smola, editors, *Advances in Kernel Methods: Support Vector Learning*. MIT Press, Cambridge, MA 1999.

C.-C. Chang and C.-J. Lin. LIBSVM: a library for support vector machines, 2001. Software available at http://www.csie.ntu.edu.tw/~cjlin/libsvm.

B. Cho, Y. Suh, S. Dieterich, and P. J. Keall. A monoscopic method for real-time tumour tracking using combined occasional x-ray imaging and continuous respiratory monitoring. *Physics in Medicine and Biology*, 53(11):2837–2855, 2008. DOI: 10.1088/0031-9155/53/11/006.

P. Demartines and J. Herault. Curvilinear component analysis: a self-organizing neural networkfor nonlinear mapping of data sets. *IEEE Transactions on Neural Networks*, 8(1):148–154, 1997. DOI: 10.1109/72.554199.

D. F. DeMenthon and L. S. Davis. Model-based object pose in 25 lines of code. *International Journal of Computer Vision*, 15(1–2):123–141, 1995. DOI: 10.1007/bf01450852.

H. Drucker, C. J. C. Burges, L. Kaufman, A. J. Smola, and V. Vapnik. Support vector regression machines, pp. 155–161 in *Advances in Neural Information Processing Systems*, volume 9 of *NIPS*, pp. 155–161. MIT Press, Cambridge, MA 1997.

F. Ernst and A. Schweikard. A family of linear algorithms for the prediction of respiratory motion in image-guided radiotherapy. In *Proceedings of the 22nd International Conference and Exhibition on Computer Assisted Radiology and Surgery (CARS'08)*, volume 3(supp. 1) of *International Journal of CARS*, pp. 31–32, Barcelona, Spain, 2008a. CARS. DOI: 10.1007/s11548-008-0169-x.

F. Ernst and A. Schweikard. Predicting respiratory motion signals for image-guided radiotherapy using multi-step linear methods (MULIN). *International Journal of Computer Assisted Radiology and Surgery*, 3(1–2):85–90, 2008b. DOI: 10.1007/s11548-008-0211-z.

F. Ernst and A. Schweikard. Prediction of respiratory motion using a modified Recursive Least Squares algorithm. In D. Bartz, S. Bohn, J. Hoffmann, editors, *7. Jahrestagung der Deutschen Gesellschaft für Computer- und Roboterassistierte Chirurgie*, pp. 157–160, Leipzig, Germany, 2008.

F. Ernst and A. Schweikard. Predicting respiratory motion signals using accurate online support vector regression (SVRpred). In *Proceedings of the 23rd International Conference and Exhibition on Computer Assisted Radiology and Surgery (CARS'09)*, volume 4(supp. 1) of *International Journal of CARS*, pp. 255–256, Berlin, Germany, 2009. CARS. DOI: 10.1007/s11548-009-0340-z.

F. Ernst and A. Schweikard. Forecasting respiratory motion with accurate online support vector regression (SVRpred). *International Journal of Computer Assisted Radiology and Surgery*, 4(5):439–447, 2009. DOI: 10.1007/s11548-009-0355-5.

F. Ernst, A. Schlaefer, and A. Schweikard. Prediction of respiratory motion with wavelet-based multiscale autoregression. In N. Ayache, S. Ourselin, and A. Maeder, editors, *MICCAI 2007, Part II*, volume 4792 of *Lecture Notes in Computer Science*, pp. 668–675, Brisbane, Australia, 2007. MICCAI, Springer. DOI: 10.1007/978-3-540-75759-7_81.

F. Ernst, A. Schlaefer, S. Dieterich, and A. Schweikard. A fast lane approach to LMS prediction of respiratory motion signals. *Biomedical Signal Processing and Control*, 3(4):291–299, 2008. DOI: 10.1016/j.bspc.2008.06.001.

F. Ernst, V. Martens, S. Schlichting, A. Beširević, M. Kleemann, C. Koch, D. Petersen, and A. Schweikard. Correlating chest surface motion to motion of the liver using ε-SVR – a porcine study. In G.-Z. Yang, D. J. Hawkes, D. Rueckert, A. Noble, and C. Taylor, editors, *MICCAI 2009, Part II*, volume 5762 of *Lecture Notes in Computer Science*, pp. 356–364, London, 2009. MICCAI, Springer. DOI: 10.1007/978-3-642-04271-3_44.

F. Ernst, A. Schlaefer, and A. Schweikard. Processing of respiratory motion traces for motion-compensated radiotherapy. *Medical Physics*, 37(1):282–294, 2010a. DOI: 10.118/1.3271684.

F. Ernst, B. Stender, A. Schlaefer, and A. Schweikard. Using ECG in motion prediction for radiosurgery of the beating heart. G.-Z. Yang and A. Darzi, editors, *Proceedings of the Third Hamlyn Symposium on Medical Robotics*, pp. 37–38. The Royal Society, London, 2010b.

R. George, S. S. Vedam, T. D. Chung, V. Ramakrishnan, and P. J. Keall. The application of the sinusoidal model to lung cancer patient respiratory motion. *Medical Physics*, 32(9):2850–2861, 2005. DOI: 10.1118/1.2001220.

D. P. Gierga, J. Brewer, G. C. Sharp, M. Betke, C. G. Willett, and G. T. Y. Chen. The correlation between internal and external markers for abdominal tumors: Implications for respiratory gating. *International Journal of Radiation Oncology, Biology, Physics*, 61(5):1551–1558, 2005. DOI: 10.1016/j.ijrobp.2004.12.013.

J. H. Goodband, O. C. L. Haas, and J. A. Mills. A comparison of neural network approaches for on-line prediction in IGRT. *Medical Physics*, 35(3):1113–1122, 2008. DOI: 10.1118/1.2836416.

J. D. P. Hoisak, K. E. Sixel, R. Tirona, P. C. F. Cheung, and J.-P. Pignol. Correlation of lung tumor motion with external surrogate indicators of respiration. *International Journal of Radiation Oncology, Biology, Physics*, 60(4):1298–1306, 2004. DOI: 10.1016/j.ijrobp.2004.07.681.

J. D. P. Hoisak, K. E. Sixel, R. Tirona, P. C. F. Cheung, and J.-P. Pignol. Prediction of lung tumour position based on spirometry and on abdominal displacement: Accuracy and reproducibility. *Radiotherapy and Oncology*, 78(3):339–346, 2006. DOI: 10.1016/j.radonc.2006.01.008.

M. Kakar, H. Nyström, L. R. Aarup, T. J. Nøttrup, and D. R. Olsen. Respiratory motion prediction by using the adaptive neuro fuzzy inference system (ANFIS). *Physics in Medicine and Biology*, 50:4721–4728, 2005. DOI: 10.1088/0031–9155/50/19/020.

E. Kanoulas, J. A. Aslam, G. C. Sharp, R. I. Berbeco, S. Nishioka, H. Shirato, and S. B. Jiang. Derivation of the tumor position from external respiratory surrogates with periodical updating of the internal/external correlation. *Physics in Medicine and Biology*, 52(17):5443–5456, 2007. DOI: 10.1088/0031–9155/52/17/023.

A. Khamene, J. K. Warzelhan, S. Vogt, D. Elgort, C. Chefd'Hotel, J. L. Duerk, J. Lewin, F. K. Wacker, and F. Sauer. Characterization of internal organ motion using skin marker positions. In C. Barillot, D. R. Haynor, and P. Hellier, editors, *MICCAI 2004, Part II*, volume 3217 of *LNCS*, pp. 526–533, St. Malo, France, 2004. MICCAI, Springer.

M. Knöpke and F. Ernst. Flexible Markergeometrien zur Erfassung von Atmungs- und Herzbewegungen an der Körperoberfläche, pp. 15–16 in D. Bartz, S. Bohn, J. Hoffmann, editors, *7. Jahrestagung der Deutschen Gesellschaft für Computer- und Roboterassistierte Chirurgie* Leipzig, Germany, 2008.

N. Koch, H. H. Liu, G. Starkschall, M. Jacobson, K. M. Forster, Z. Liao, R. Komaki, and C. W. Stevens. Evaluation of internal lung motion for respiratory-gated radiotherapy using MRI: Part I–correlating internal lung motion with skin fiducial motion. *International Journal of Radiation Oncology, Biology, Physics*, 60(5):1459–1472, 2004. DOI: 10.1016/j.ijrobp.2004.05.055.

J. A. Lee, A. Lendasse, and M. Verleysen. Curvilinear distance analysis versus isomap. In *European Symposium on Artificial Neural Networks (ESANN)*, pp. 185–192, Bruges (Belgium), 2002.

J. Ma, J. Theiler, and S. Perkins. Accurate on-line support vector regression. *Neural Computation*, 15(11):2683–2703, 2003. DOI: 10.1162/089976603322385117.

K. C. McCall and R. Jeraj. Dual-component model of respiratory motion based on the periodic autoregressive moving average (periodic ARMA) method. *Physics in Medicine and Biology*, 52(12):3455–3466, 2007. DOI: 10.1088/0031–9155/52/12/009.

J. R. McClelland, J. M. Blackall, S. Tarte, A. C. Chandler, S. Hughes, S. Ahmad, D. B. Landau, and D. J. Hawkes. A continuous 4D motion model from multiple respiratory cycles for use in lung radiotherapy. *Medical Physics*, 33(9):3348–3358, 2006. DOI: 10.1118/1.2222079.

M. J. Murphy. Using neural networks to predict breathing motion. In *7th International Conference on Machine Learning and Applications*, pp. 528–532, Los Alamitos, CA, USA, 2008. IEEE Computer Society. DOI: 10.1109/icmla.2008.136.

M. J. Murphy and S. Dieterich. Comparative performance of linear and nonlinear neural networks to predict irregular breathing. *Physics in Medicine and Biology*, 51(22):5903–5914, 2006. DOI: 10.1088/0031–9155/51/22/012.

M. J. Murphy and D. Pokhrel. Optimization of an adaptive neural network to predict breathing. *Medical Physics*, 36(1):40–47, 2009. DOI: 10.1118/1.3026608.

M. J. Murphy, M. Isaaksson, and J. Jaldén. Adaptive filtering to predict lung tumor breathing motion during imageguided radiation therapy. In *Proceedings of the 16th International Conference and Exhibition on Computer Assisted Radiology and Surgery (CARS'02)*, volume 16, pp. 539–544, Paris, France, 2002.

F. Parrella. Online support vector regression. Master's thesis, University of Genoa, 2007.

D. Putra, O. C. L. Haas, J. A. Mills, and K. J. Burnham. A multiple model approach to respiratory motion prediction for real-time IGRT. *Physics in Medicine and Biology*, 53(6):1651–1663, 2008. DOI: 10.1088/0031–9155/53/6/010.

L. Ramrath, A. Schlaefer, F. Ernst, S. Dieterich, and A. Schweikard. Prediction of respiratory motion with a multi-frequency based Extended Kalman Filter. In *Proceedings of the 21st International Conference and Exhibition on Computer Assisted Radiology and Surgery (CARS'07)*, volume 2(supp. 1) of *International Journal of CARS*, pp. 56–58, Berlin, Germany, 2007. CARS. DOI: 10.1007/s11548-007-0083-7.

Q. Ren, S. Nishioka, H. Shirato, and R. I. Berbeco. Adaptive prediction of respiratory motion for motion compensation radiotherapy. *Physics in Medicine and Biology*, 52(22):6651–6661, 2007. DOI: 10.1088/0031–9155/52/22/007.

C. N. Riviere, A. Thakral, I. I. Iordachita, G. Mitroi, and D. Stoianovici. Predicting respiratory motion for active canceling during percutaneous needle insertion. In *Proceedings of the 23rd Annual International Conference of the IEEE Engineering in Medicine and Biology Society*, volume 4, pp. 3477–3480, 2001.

D. Ruan, J. A. Fessler, and J. M. Balter. Real-time prediction of respiratory motion based on local regression methods. *Physics in Medicine and Biology*, 52(23):7137–7152, 2007. DOI: 10.1088/0031-9155/52/23/024.

N. Rzezovski and F. Ernst. Graphical tool for the prediction of respiratory motion signals. pp. 179–180 in D. Bartz, S. Bohn, J. Hoffmann, editors, *7. Jahrestagung der Deutschen Gesellschaft für Computer- und Roboterassistierte Chirurgie* Leipzig, Germany, 2008.

A. Sahih, O. C. L. Haas, J. H. Goodband, D. Putra, J. A. Mills, and K. J. Burnham. Respiratory motion prediction for adaptive radiotherapy. In *IAR – ACD 2006*, Nancy, France, 2006. German-French Institute for Automation and Robotics.

S. Sayeh, J. Wang, W. T. Main, W. Kilby, and C. R. Maurer Jr. Respiratory motion tracking for robotic radiosurgery, pp. 15–30, In *Robotic Radiosurgery. Treating Tumors that Move with Respiration*. In uklt_07, 1st edition, 2007. DOI: 10.1007/978-3-540-69886-9.

A. Schweikard, G. Glosser, M. Bodduluri, M. J. Murphy, and J. R. Adler, Jr. Robotic Motion Compensation for Respiratory Movement during Radiosurgery. *Journal of Computer-Aided Surgery*, 5(4):263–277, 2000. DOI: 10.3109/10929080009148894.

A. Schweikard, H. Shiomi, and J. R. Adler, Jr. Respiration tracking in radiosurgery. *Medical Physics*, 31(10):2738–2741, 2004. DOI: 10.1118/1.1774132.

Y. Seppenwoolde, R. I. Berbeco, S. Nishioka, H. Shirato, and B. Heijmen. Accuracy of tumor motion compensation algorithm from a robotic respiratory tracking system: A simulation study. *Medical Physics*, 34(7):2774–2784, 2007. DOI: 10.1118/1.2739811.

G. C. Sharp, S. B. Jiang, S. Shimizu, and H. Shirato. Prediction of respiratory tumour motion for real-time image-guided radiotherapy. *Physics in Medicine and Biology*, 49(3):425–440, 2004. DOI: 10.1088/0031-9155/49/3/006.

Y. Sheng, S. Li, S. Sayeh, J. Wang, and H. Wang. Fuzzy and hybrid prediction of position signal in synchrony respiratory track-ing system. pp. 459–464 in R. J. P. de Figueiredo, editor, *SIP 2007*, Honolulu, USA, 2007. IASTED, Acta Press.

A. J. Smola and B. Schölkopf. A tutorial on support vector regression. *Statistics and Computing*, 14:199–222, 2004.

J. Tang, S. Dieterich, and K. Cleary. Respiratory motion tracking of skin and liver in swine for CyberKnife motion compensation. In R. L. Galloway, Jr., editor, *Medical Imaging 2004: Visualization, Image-Guided Procedures, and Display*, volume 5367(1), pp. 729–734, San Diego, CA, USA, 2004. SPIE. DOI: 10.1117/12.536941.

J. B. Tenenbaum, V. de Silva, and J. C. Langford. A global geometric framework for nonlinear dimensionality reduction. *Science*, 290:2319–2323, 2000.

H. C. Urschel, Jr., J. J. Kresl, J. D. Luketich, L. Papiez, and R. D. Timmerman, editors. *Robotic Radiosurgery. Treating Tumors that Move with Respiration*. Springer, Berlin, 1st edition, 2007. DOI: 10.1007/978-3-540-69886-9.

S. S. Vedam, V. R. Kini, P. J. Keall, V. Ramakrishnan, H. Mostafavi, and R. Mohan. Quantifying the predictability of diaphragm motion during respiration with a noninvasive external marker. *Medical Physics*, 30(4):505–513, 2003. DOI: 10.1118/1.1558675.

S. S. Vedam, P. J. Keall, A. Docef, D. A. Todor, V. R. Kini, and R. Mohan. Predicting respiratory motion for four-dimensional radiotherapy. *Medical Physics*, 31(8):2274–2283, 2004. DOI: 10.1118/1.1771931.

M. Verleysen and D. Francois. The curse of dimensionality in data mining and time series prediction. In *Proceedings of the 8th International Workshop on Artificial Neural Networks (IWANN 2005)*, volume 3512 of *Lecture Notes in Computer Science*, pp. 758–770, Barcelona, Spain, 2005. Springer. DOI: 10.1007/11494669_93.

J. B. West. *Respiratory Physiology: The Essentials*. Lippincott Williams & Wilkins, Philadelphia, PA 8 edition, 2008.

H. Yan, F.-F. Yin, G.-P. Zhu, M. Ajlouni, and J. H. Kim. Adaptive prediction of internal target motion using external marker motion: a technical study. *Physics in Medicine and Biology*, 51(1):31–44, 2006. DOI: 10.1088/0031-9155/51/1/003.

8

Treatment Planning for Motion Adaptation in Radiation Therapy

Alexander Schlaefer

8.1 Introduction 65
8.2 Treatment Planning in Radiation Therapy 65
8.3 Image Processing in Treatment Planning 66
8.4 Planning for Motion-Compensated Treatment 69
8.5 Intratreatment Motion Adaption during Treatment Planning 72
References 73

8.1 Introduction

The core objective of treatment planning is to find treatment parameters that maximize the treatment effect in the target region while reducing potential complications in normal tissue. Intensity modulation and noncoplanar beam arrangements allow for unprecedented flexibility in shaping the dose distribution. Moreover, sophisticated pretreatment imaging and image-guided treatment can provide information about shape, motion, and deformation of internal structures, and this information needs to be reflected in the planning process.

We briefly introduce planning as a multicriteria problem and focus on the aspects of organ motion and plan adaption. A key motivation for motion compensation is the typically much smaller high-dose region and the potential for better sparing of normal tissue. We illustrate how 4D planning can be used to incorporate motion and how this extends to intrafraction adaption of the treatment plan. With motion compensation, organ movements due to respiration are not considered a random error. The systems used to estimate and track the motion and to deliver the beams synchronously can substantially reduce the uncertainty with respect to the target dose. However, it is also important to understand the limitations of 4D CT, tracking, and motion compensation, and to include the remaining uncertainty when creating a treatment plan.

8.2 Treatment Planning in Radiation Therapy

One of the key advantages of radiation therapy is its noninvasive nature. However, the use of radiation implies the need for a careful balance between conflicting clinical goals, notably the dose in the target and the sparing of critical structures and normal tissue. Treatment planning attempts to establish the best trade-off with respect to criteria such as target coverage, conformality and homogeneity of the dose distribution, total treatment time, and dose to critical structures. Hence, planning is based on a definition of the target and all relevant critical structures, a set of clinical goals, and a notion of what constitutes a preferable balance among these goals. The resulting treatment plan specifies the parameters for the respective delivery system. For now, we abstract from the actual beam delivery details and consider a plan as a set of beams, where each beam is described by its shape, orientation, and beam-on time. Note that these beams could overlap to realize intensity-modulated radiation therapy.

In principle, it is possible to create treatment plans by forward planning, for example, specifying a set of isocentric and equally weighted beams. Such plans may be sufficient for simple target shapes like treating an isolated spherical lung lesion using a large number of beam directions with the CyberKnife®. However, typically, some sort of intensity modulation will be required to fulfill the desired clinical goals, and inverse planning (Censor et al. 1988; Webb 1989; Bortfeld et al. 1990) is used to search for a treatment plan that represents a clinically acceptable balance between the planning goals. Different optimization methods have been proposed for this search, including gradient decent, simulated annealing, and linear or quadratic programming. Generally, these methods require a mapping of the clinical goals to some mathematical objective function, for example, measuring the deviation from the desired dose thresholds. Note that while the resulting treatment plan is usually optimal with respect to the mathematical function, it may not represent the best compromise regarding the clinical goals. Conventionally, this leads to an iterative planning process, where the weight of the clinical goals in the objective function is modified to reflect their importance to the planner and then a new plan is computed. One limitation of this approach is that the relative importance of the different goals does not necessarily map to a respective ratio for the weights. In fact, it may be hard to see whether any setting of the weights will yield the desired balance between the clinical goals.

More recently it has been appreciated that treatment planning resembles a multicriteria optimization problem (Cotrutz et al. 2001; Hamacher and Küfer 2002). Given the conflicting nature of the clinical goals, any plan represents a trade-off with respect to the planning criteria. Clearly, the optimal plan cannot be improved with respect to any one criterion without compromising other criteria, and methods to search for plans fulfilling this requirement have been introduced to treatment planning (Romeijn et al. 2004; Schreibmann et al. 2004; Craft et al. 2005; Jee et al. 2007; Wilkens et al. 2007; Schlaefer and Schweikard 2008). Figure 8.1 illustrates an example for a plan generated in a stepwise fashion, addressing one criterion at a time.

First, the coverage of the target is maximized and maintained with constraints (Figure 8.1a). Second, the dose in the rectum is reduced (Figure 8.1b). Note that the target dose did not change. Finally, after fixing the rectal dose with a constraint, further sparing of the bladder is studied (Figure 8.1c). Clearly, to fine-tune the balance, constraints on the dose distribution can be relaxed at any time. However, the possible improvements with respect to the corresponding criteria are always established by optimization, that is, highlighting the best possible solution for the given constraints. Figure 8.1 also shows that the optimal set of beams and the beam-on times depend on the desired trade-off. The advantage of this stepwise approach compared with using importance factors is illustrated in Figure 8.2, where the trade-off between conformality and homogeneity for two stereotactic body radiation therapy cases is studied. The desired target dose corresponds to different ratios of the importance factors, while the same optimization steps lead to a good starting point for further fine tuning the trade-off.

In order to run the optimizations, the planning problem has to be expressed in mathematical form. This is typically done by discretizing the volumes of interest (VOI) into a finite set of voxels, and by computing the dose each beam delivers in each voxel (Censor et al. 2006). The definition of the VOI is typically a manual or semiautomatic process based on pretreatment image data. Conventionally, 3D computed tomography (CT) is widely used to delineate contours of gross target volume (GTV), clinical target volume (CTV) and the organs at risk (OAR) (Censor et al. 2006). It is worthwhile considering the underlying mathematical optimization problem when defining the VOI, as overlapping volumes will aggravate conflicts during planning, and discretization may impact the shape of the structures as seen by the optimization.

There are various sources of uncertainty related to the VOI, for example, the contouring itself, or setup errors (Ekberg et al.

1998; Van de Steene et al. 2002; Van Herk 2004). Clearly, this is also where motion and deformation directly impact planning. The standard way to consider these variations in the contours is adding a margin around CTV and OAR, leading to planning target volume (PTV) and planning organ at risk volumes (PRV). While the resulting PTV substantially reduces the impact of tumor motion on the dose delivered to the target, it will also lead to an increased dose delivered to nontarget tissue. Additionally, the planning problem itself might become harder to solve if the volume is larger and potentially closer to PRVs.

Hence, it is important to consider the type of motion, its potential effect on the treatment, and the means to detect and compensate the motion during planning. For example, spontaneous motion during the treatment of spinal lesion could cause a misalignment that will affect the dose delivery for the remaining fraction. However, if the position of the spine is tracked throughout treatment, the impact can be reduced substantially (Furweger et al. 2009). Likewise, prostate motion can be tracked and compensated for (Xie et al. 2008). While the extent of spontaneous motion is typically hard to predict, systematic motion due to respiration or pulsation can be estimated, and actively compensated. Hence, this compensation needs to be considered during planning. Another aspect is the potential for plan adaption. For typical fractionation, the errors may be compensated in later fractions. With hypofractionation or when treatment is delivered in a single fraction, adaption has to happen intrafraction, that is, online.

8.3 Image Processing in Treatment Planning

Treatment planning typically involves the acquisition of CT scans to delineate the VOI and to calculate the dose. The main difference when planning motion-compensated treatment is the focus on organ movements. Despite technical advancements in CT technology (Kalender 2006), the temporal resolution of current systems does not allow to obtain artifact-free images of organs moving due to respiration. To overcome this limitation, CT has been combined with systems measuring respiration, and 4D CT is typically computed in a postprocessing step by sorting the images according to the breathing signal (Low et al. 2003; Ford et al. 2003; Vedam et al. 2003; Pan et al. 2004). Hence, the resulting stack of 3D images incorporates data from multiple breathing cycles. In practice, the sorting is often done simply based on the respiratory phase. Given the variability of breathing and since neighboring CT slices do not perfectly represent the same breathing state, the images are prone to artifacts, particularly for phase-based binning. Including information on displacement and velocity can help improve the quality of 4D CT (Langneer and Keall 2008). Another approach is the use of MRI to estimate respiratory motion (Plathow et al. 2004; Blackall et al. 2006).

A 4D CT presents a large data set, and contouring the VOI in all slices is impractical. Moreover, treatment planning needs to

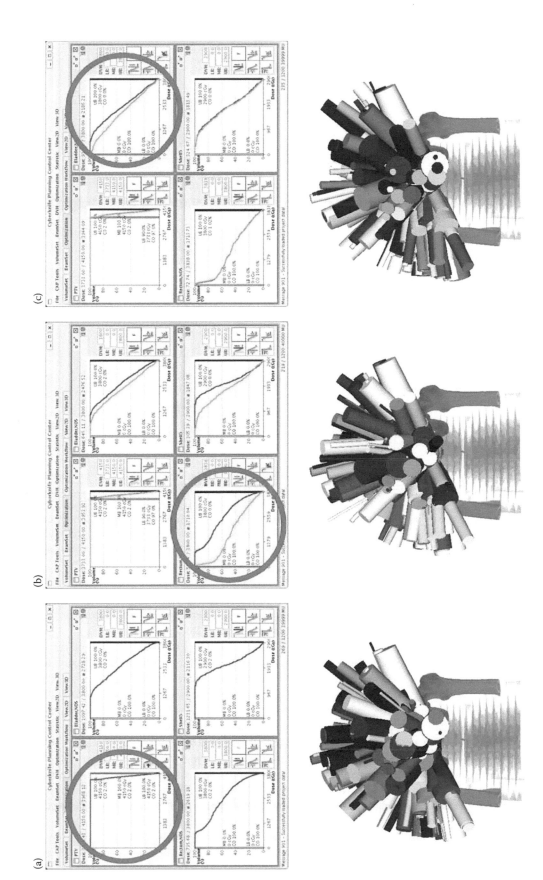

FIGURE 8.1 Stepwise improvement of a treatment plan for CyberKnife treatment of a prostate target. The top row shows the dose volume histograms, the bottom row the related beam sets with lighter shading of the beams indicating more monitor units. First, coverage in the prostate is maximized (a). Second, the dose in the rectum is minimized (b). Note the change in the DVH for the rectum (*circle*), while the prostate dose was maintained with a strict bound. Third, the dose in the bladder is minimized with a strict bound. Again, the dose in PTV and rectum do not change, as they are constrained to the results of the previous optimization steps.

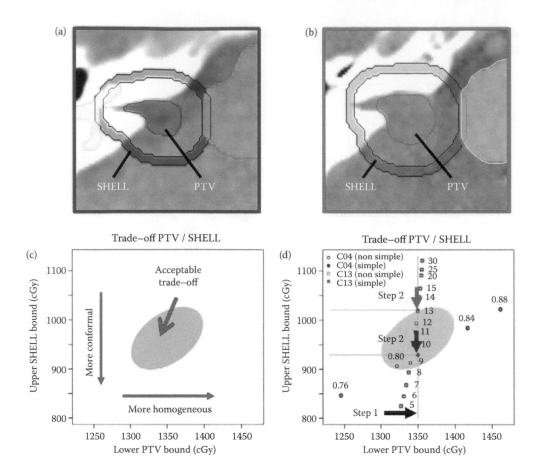

FIGURE 8.2 The trade-off between conformality and homogeneity for two stereotactic body radiation therapy cases is studied. An example illustrating stepwise planning. For simplicity, two different acoustic neuroma cases (Part figures (a) and (b)) are considered, each with the PTV, a SHELL structure, and OAR contoured. (c) shows the planning criteria, i.e., when all other criteria remain bounded, homogeneity improves with an increasing lower PTV bound and conformity improves with a decreasing upper SHELL bound. In the part figure (d), the squares and circles illustrate the difference in the importance factor ratio for the two plans, while the two steps indicated by arrows - first, optimizing the lower PTV bound, second, optimizing the upper SHELL bound - provide good initial solutions to explore the trade-off, compare. (Adapted from Schlaefer, A., Schweikard, A. *Med. Phys.*, 35(5), 2094–103, 2008.)

relate the information in the different 3D images in order to compute the total dose, and various methods for image registration has been proposed to achieve this (Guerrero et al. 2005; Foskey et al. 2005; Schreibmann et al. 2006; McClelland et al. 2006; Ehrhardt et al. 2007; Yang et al. 2008; Schreibmann et al. 2008). Typically, the contouring is done on a 3D image representing one breathing state that is registered with the 3D images for all remaining breathing states. For example, the deformation between two 3D image A and B can be expressed using B-Splines, which map each point in image A to the corresponding point in image B. Clearly, the images can also be registered against a different 3D CT, or with MRI data. For example, a conventional 3D CT acquired in breath hold may show fewer artifacts and therefore be better suited for contouring, while the registration with the 4D CT images provides information on the organ motion. It is worth noting that image registration is itself an optimization problem, that is, the algorithms search for a mapping that minimizes an objective function. This objective measures the distance in terms of image features, but it often

also includes a smoothing term to handle artifacts. Hence, the quality of the treatment plan will also depend on the registration method (Brock et al. 2010).

Most image guidance is based on X-ray images, and since few tumors are visible in the radiographs, typically, small fiducials are implanted as artificial landmarks. These fiducials are placed before the planning image is acquired and can also be used as landmarks for image registration. For motion-compensated treatment, planning can also include criteria related to the tracking of motion. Tracking requires the fiducials to be visible in the intratreatment imaging. When the treatment is delivered with the CyberKnife, the workspace of the robot can be constrained such that the X-ray cameras remain unblocked. For MLC-based motion compensation, the MV images can be used to locate the target position. Hence, it is preferable to select MLC segments that leave the fiducial markers visible in the MV images (Ma et al. 2009). Note that this does not necessarily mean a substantial degradation of plan quality with respect to clinical criteria. The planning problem is typically degenerate (Alber et al. 2002), that is, there are multiple sets of beams and beam weights that lead to similar plans.

8.4 Planning for Motion-Compensated Treatment

Conventionally, organ motion is considered a source of error and handled by margins (Van Herk 2004). Following ICRU 62, the CTV is extended by the internal margin (IM) accounting for organ motion and deformation to form the internal target volume (ITV). The size of the margin is either based on typical motion patterns for the respective target organ, or estimated from imaging. Using 4D CT, the size and position of the VOIs can be estimated for different breathing states. It would be possible to consider the maximum extent of CTV motion when defining the IM, for example, computing the hull over the CTV in different breathing states. Figure 8.3a and b illustrate this approach. Note that in the example, the CTV is moving and also rotating and expanding.

However, if the motion pattern can be expressed with a probability distribution, this information can be included into planning (Lujan et al. 1999; Yan et al. 1999,;Mageras et al. 1999; Li and Xing 2000; Shirato et al. 2000; Zhang et al. 2004; Unkelbach and Oelfke 2004). Consider a voxel v, and a probability distribution P(v, x) describing the likelihood that the center of voxel v is point x. Now, the expected dose delivered by a beam b in voxel v is the integral of the dose delivered in all points p weighted by the respective probability P(v, p). For respiratory motion, the position and shape of the CTV in different breathing states can be estimated, for example, from 4D CT. For 4D CT binned according to the respiratory phase, the probability of being in one phase is simply 1 over the number of phases. Instead of computing the dose based on a single 3D CT scan, the dose is calculated for each respiratory phase from the 4D CT and then accumulated over all phases.

When the respiratory motion of the CTV is incorporated in the dose calculation, the resulting dose distribution accounts for this motion. This can be seen as computing the appropriate margin, which does not necessarily cover the whole CTV during in all breathing states. Figure 8.3c illustrates that the CTV is subjected to different doses during a respiratory cycle. The gray, dark-gray, and light-gray shading corresponds to a dose at, above, and below the prescribed dose, respectively. As the dose delivered in the different positions adds up, the lower doses in breathing states on the right will be compensated by the higher dose in breathing states in the center. Hence, the actual region that needs to receive a dose above the prescribed dose does not encompass the CTV in all individual breathing states, as shown in Figure 8.3d. Clearly, the shape of this region depends on the dose distribution and is not known before 4D inverse planning. We indicate this by the acronym 4D-ITV, noting that a similar approach can be used to obtain the PRVs.

While we have reduced the uncertainty with respect to motion and deformation of the VOI, we have introduced new sources of potential error. The acquisition of 4D CT itself is subject to artifacts. Moreover, the breathing pattern may change over time, and the assumptions regarding position and shape of the VOI may no longer hold. Finally, even when 4D planning is employed, the high-dose region is typically much larger than the CTV in any breathing state. The latter issue is the key motivation for active motion compensation. If beams and CTV move perfectly synchronous, the uncertainty due to changes in the CTV position is zero. However, Figure 8.4a–c illustrate that deformations and rotations still need to be considered for the ITV. The shape and orientation of the CTV in different respiratory phases can be estimated from 4D CT. Current clinical motion compensation is mostly based on tracking small radio-opaque fiducial markers. These markers are placed close to the CTV and easy to detect in the CT images, and they can be used for registering the CTV from different respiratory phases into the same coordinate frame. Often, the 4D CT phases corresponding to inhalation and exhalation will define most of the resulting ITV shape.

Of course, it is possible to incorporate the variations in shape and rotation and the expected beam motion in the dose calculations and to compute the actual high-dose region as a result of 4D inverse planning (Keall et al. 2005; Schlaefer et al. 2005). While the difference between ITV and 4D-ITV is typically smaller than that with static beams, Figure 8.4c–e illustrates that the same argument regarding the aggregation of the dose applies. Particularly for stereotactic treatments, the dose distribution is often highly conformal, but less homogeneous. If the CTV expands and shrinks over the course of breathing, voxels on its surface will move into higher dose when the volume is smaller and into lower dose when the volume is larger. Consider the left side of Figure 8.4c, where the volume is small and covered by a dose above the prescribed dose (dark gray). In the respiratory states shown in the center of Figure 8.4c, the surface voxels get just the prescribed dose (gray). Only in the state shown on

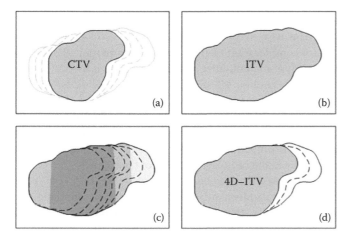

FIGURE 8.3 (a) The CTV moves due to respiration, that is, its position, shape and orientation change; (b) a simple approach to create the ITV is to generate the hull over the CTV in all breathing states; (c) the delivered dose does not need to be completely homogeneous, that is, dose above the prescribed dose (*dark gray*) can make up for dose below the prescribed dose (*light gray*); (d) the actual volume receiving a dose above the prescribed dose is computed by inverse planning.

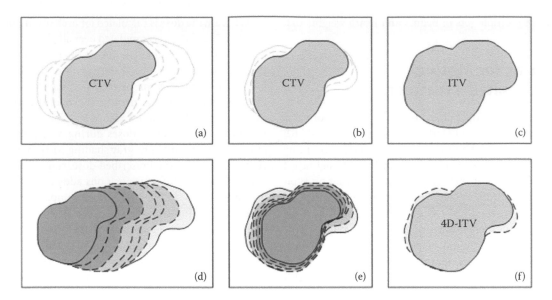

FIGURE 8.4   (a) The CTV moves due to respiration, that is, its position, shape, and orientation change; (b) if motion is compensated the CTVs in different breathing states are registered and only deformations and rotations need to be considered; (c) the hull over the registered CTVs is a simple estimate for the ITV; (d) the delivered dose is typically not completely homogeneous, that is, dose above the prescribed dose (*dark gray*) can make up for dose below the prescribed dose (*light gray*); (e) the delivered dose depends on the breathing state, for a conformal dose distribution the surface will often get higher dose when the volume contracts, and lower dose when it expands; (f) the actual volume receiving a dose above the prescribed dose is computed by inverse planning, note that this considers motion compensation

the left side, the surface voxels get a dose below the prescribed dose (light gray). Figure 8.4d illustrates how this corresponds to the steep gradient at the target's surface, and in Figure 8.4e, the resulting 4D-ITV is shown.

Comparison of Figures 8.3d and 8.4e demonstrates how the ITV can be substantially smaller when active motion compensation is used to treat targets with large respiratory motion. However, while the beams move synchronously with the CTV, other structures may exhibit different motion patterns. Hence, the beams move relative to these structures, Consider the motion pattern and geometry shown in Figure 8.5a; here, the gray ellipse represents a mobile target moving up and down, and the circle denotes a static critical structure. To illustrate the potential effect of beam motion, we assume planning at inhalation. As the beams move with the target to its exhalation position, some beams may move into the critical structure. Figure 8.5b–e shows dose distributions for four different scenarios. In Figure 8.5b, the dose distribution after the 3D planning is shown, with the red isodose denoting the prescription dose and the blue isodose denoting the maximum dose for the critical structure. If this plan was delivered without motion compensation, the target would partially move out of the high-dose region (Figure 8.5c). For a motion-compensated treatment, the target will get the prescribed dose (Figure 8.5d). However, the isodose for the maximum dose in the critical structure is shifted, violating this constraint. This can be addressed by computing a 4D plan that takes the motion of the VOI and the beams into account. Figure 8.5e shows that the isodose lines for target and critical structure fulfill the planning constraints.

The example from Figure 8.5a–e corresponds to a treatment in lung or abdomen, in which the target shows noticeable anterior–posterior motion and the spinal cord is an immobile critical structure. For many typical stereotactic treatments, the steep and conformal dose gradient and the blurring effect of the motion will lead to acceptable doses for the spinal cord, even if the beams move. This is also illustrated in Figure 8.6, in which the fraction of the critical structure that is overdosed is rather small. However, the example was simulating a CyberKnife-like treatment with many beams from a large solid angle. A more careful consideration of the relative beam motion is advisable when only a small number of beam directions is used.

In our example, 4D planning can be seen as an implicit computation of the ITV and the motion margin for the OARs. Furthermore, a careful inspection of the dose–volume histogram in Figure 8.6 shows a very small underdosing of the target for motion-compensated treatment using a 3D plan. This is due to changes in the radiological depth of the target.

As it is based on dose calculations using the 4D CT, 4D planning will also address this effect. Of course, the limitations of 4D CT still hold, that is, imaging artifacts and changes in the breathing pattern add uncertainty. Note that the effect of different depth and tissue density would be much more severe for particle therapy (Bert and Rietzel 2007).

Moreover, active motion compensation introduces new sources of uncertainty. First, the target position needs to be established. For fiducial-based tracking, the error due to image processing is typically below 1 mm (Hatipoglu et al. 2007; Muacevic et al. 2007; Seppenwoolde et al. 2007; Hoogeman et al. 2009), and similar

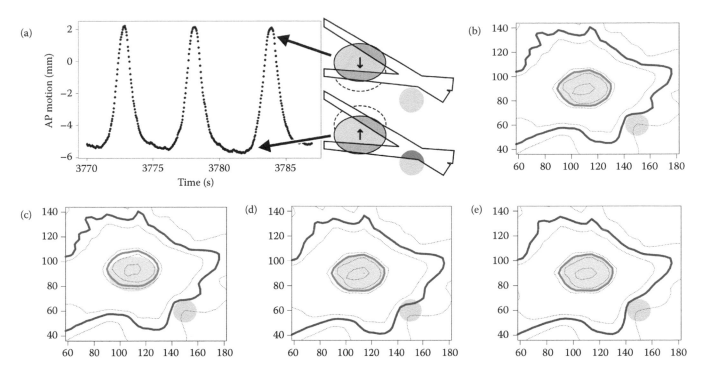

FIGURE 8.5 **(See color insert.)** (a) a respiratory motion pattern and the position of beams and VOI during inhalation (planning state) and exhalation; (b–e) dose distributions after planning, for no motion compensation, for motion compensation with 3D planning, and for motion compensation with 4D planning, respectively.

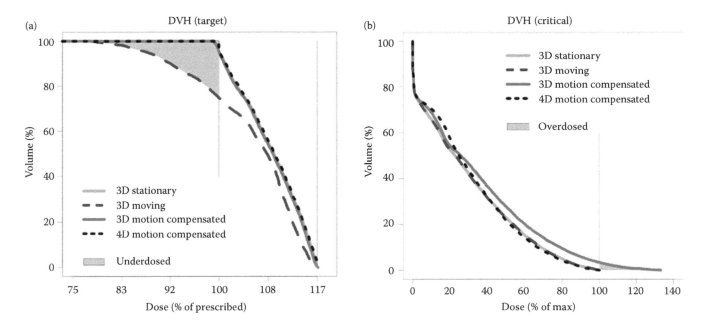

FIGURE 8.6 **(See color insert.)** The dose distribution in target (a) and critical structure (b). Uncompensated target motion would lead to severe underdosing, and for motion compensation with 3D planning a very small deviation in the dose is cause by differences in the beam path between dose computation and moving beam. A more noticeable overdosing of the critical structure can be observed. Both deviations can be handled by 4D planning

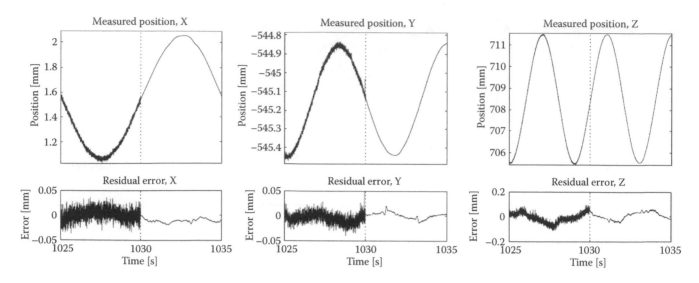

FIGURE 8.7 Tracking data for a chest marker before and after smoothing. The three plots show *x*, *y*, and *z* positions of a known sinusoidal trajectory and the estimated tracking error.

errors can be expected when direct tracking of the tumor is possible (Cui et al. 2007; Fu et al. 2007). Second, current approaches to active tracing require a surrogate signal to avoid excessive imaging dose. Optical tracking of external markers placed on the patient's chest has been proven to be a practical approach. Even though the absolute error of typical tracking systems is small (Ernst et al. 2010), it will impact the overall tracking error, particularly as the relatively small motion of the chest is correlated to potentially much larger target motion. Figure 8.7 shows how filtering of the raw measurements can substantially reduce the noise. However, the remaining uncertainty and a potential variation in the correlation model in between two image-verified target positions need to be considered during planning. Finally, to overcome the system latency and move the beams synchronous to the target, prediction of its motion is required (Sharp et al. 2004; Murphy and Dieterich 2006; Ernst et al. 2008; Putra et al. 2008). The residual error is small but adds to the overall uncertainty.

8.5 Intratreatment Motion Adaption during Treatment Planning

In contrast to conventional radiotherapy, active motion compensation can provide feedback on the actual tracking error during treatment. For example, the imaging system of the CyberKnife establishes the position of the fiducial marker in specific intervals. The predicted position of the target can be compared with the actual position to estimate the tracking error. So far, this information is primarily used to monitor the treatment and to restart tracking when the error grows beyond a threshold. Clearly, if variations from the planning assumptions are detected, it would be reasonable to adapt the plan during treatment.

For example, changes in the motion pattern would directly impact 4D planning. Some investigations indicate that such variations can be substantial (Juhler-Nottrup et al. 2007; Von

Siebenthal et al. 2007; Minn et al. 2009). Moreover, changes in the breathing pattern seem more likely during single- or hypofractionated treatment, where delivery of a treatment fraction can take more than thirty minutes. While coaching the patient can mitigate the problem, it may not be feasible for all patients, for example, elderly patients or patients with complications from lung cancer. Another approach is to compute a new plan based on an estimate of the dose that has been delivered and the new respiratory pattern. Figure 8.8 shows the motion of a pancreatic tumor extracted from the CyberKnife log files. The two motion patterns are taken from the same treatment fraction, just two minutes apart. Clearly, the treatment plans will look different depending on the motion pattern assumed for planning. This is illustrated in Figure 8.9, which shows three different CyberKnife treatment plans corresponding to 3D planning (a), 4D planning with the black motion pattern (b), and 4D planning with the gray motion pattern (c). The beam-on time is represented by the beams' color, that is, beams with a lighter shade have longer beam-on times. There are noticeable differences in the three plans, mostly in the regions highlighted by the arrows.

Intratreatment adaption of the plan poses a number of challenges. First, changes in the breathing pattern need to be detected and predicted. Second, the plan optimization must be sufficiently fast and automatic to compute a new plan while a beam is delivered. Advances in fast computing hardware already allow for rapid adaption of treatment plans (Men et al. 2009). However, the trade-off among the planning criteria may shift during automatic adaption, and criteria to decide when adaption is beneficial should be included. Finally, the actual information on motion and deformation must be acquired. Currently, in most cases, this is limited to rather infrequent measurements of fiducial positions. In order to detect and address deformations during motion-compensated treatment, fast 4D imaging is required. Although 4D cone-beam CT can be acquired immediately before the treatment starts (Sonke et al. 2005), it cannot be

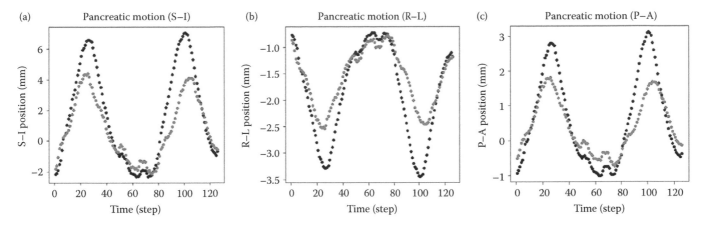

FIGURE 8.8 The superior-inferior (a), right-left (b), and posterior-anterior (c) components of intrafraction pancreatic tumor motion, initially (black) and approximately two minutes later (gray).

FIGURE 8.9 Three different Cyberknife treatment plans for a pancreatic tumor based on (a) 3D planning; (b) 4D planning and the initial motion pattern from Figure 8.8; (c) 4D planning and the second motion pattern from Figure 8.8. Arrows highlight area with most differences.

used to monitor motion during treatment. Systems integrating MRI with the treatment delivery are currently under development and could provide improved motion monitoring. Given the improvements in 4D ultrasound, this may provide another alternative, particularly when cardiac motion needs to be considered.

References

M. Alber, G. Meedt, F. Nüsslin, and R. Reemtsen. On the degeneracy of the IMRT optimization problem. *Med. Phys.*, 29(11):2584–2589, 2002.

C. Bert and E. Rietzel. 4D treatment planning for scanned ion beams. *Radiat. Oncol.*, 2:24, 2007.

J. M. Blackall, S. Ahmad, M. E. Miquel, J. R. McClelland, D. B. Landau, and D. J. Hawkes. MRI-based measurements of respiratory motion variability and assessment of imaging strategies for radiotherapy planning. *Phys. Med. Biol.*, 51, 4147–4169, 2006.

T. Bortfeld, J. Bürkelbach, R. Boesecke, and W. Schlegel. Methods of image reconstruction from projections applied to conformation radiotherapy. *Phys. Med. Biol.*, 35, 1423–34, 1990.

K. K. Brock. Deformable Registration Accuracy Consortium: Results of a multi-institution deformable registration accuracy study (MIDRAS). *Int. J. Radiat. Oncol. Biol. Phys.*, 76(2), 583–96, 2010.

Y. Censor, M. D. Altschuler, and W. D. Powlis. A computational solution of the inverse problem in radiation-therapy treatment planning. *Appl. Math. Comput.*, 25, 57–87, 1988.

Y. Censor, T. Bortfeld, B. Martin, and A. Trofimov. A unified approach for inversion problems in intensity-modulated radiation therapy. *Phys. Med. Biol.*, 51, 2353–65, 2006.

C. Cotrutz, M. Lahanas, K. Kappas, and D. Baltas, A multiobjective gradient based dose optimization algorithm for external beam conformal radiotherapy. *Phys. Med. Biol.*, 46, 2161–75, 2001.

D. Craft, T. Halabi, and T. Bortfeld. Exploration of tradeoffs in intensity-modulated radiotherapy. *Phys. Med. Biol.*, 50, 5857–68, 2005.

Y. Cui, J. G. Dy, G. C. Sharp, B. Alexander, S. B. Jiang. Multiple template-based fluoroscopic tracking of lung tumor mass without implanted fiducial markers. *Phys. Med. Biol.*, 52, 6229–42, 2007.

J. Ehrhardt, R. Werner, D. Säring, T. Frenzel, W. Lu, D. Low, and H. Handels. An optical flow based method for improved reconstruction of 4D CT data sets acquired during free breathing. *Med. Phys.*, 34(2), 711–21, 2007.

L. Ekberg, O. Holmberg, L. Wittgren, G. Bjelkengren, and T. Landberg. What margins should be added to the clinical target volume in radiotherapy treatment planning for lung cancer? *Radiother. Oncol.*, 48, 71–77, 1998.

F. Ernst, A. Schlaefer, and A. Schweikard. Smoothing of respiratory motion traces for motion-compensated radiotherapy. *Med. Phys.*, 37(1), 282–294, 2010.

F. Ernst, A. Schlaefer, S. Dieterich, and A. Schweikard. A Fast Lane Approach to LMS Prediction of Respiratory Motion Signals. *Biomed. Signal Process Contr.*, 3(4), 291–299, 2008.

E. C. Ford, G. S. Mageras, E. Yorke, and C. C. Ling. Respiration correlated spiral CT: A method of measuring respiratory-induced anatomic motion for radiation treatment planning. *Med. Phys.* 30(1), 88–97, 2003.

M. Foskey, B. Davis, L. Goyal, S. Chang, E. Chaney, N. Strehl, S. Tomei, J. Rosenman, and S. Joshi. Large deformation three-dimensional image registration in image-guided radiation therapy. *Phys. Med. Biol.*, 50, 5869–5892, 2005.

D. Fu, R. Kahn, B. Wang, H. Wang, Z. Mu, J. Park, G. Kuduvalli, and C. R. Maurer Jr. XsightTM lung tracking: A fiducial-less method for respiratory motion tracking. In *Robotic Radiosurgery. Treating Tumors That Move with Respiration*, Springer, Berlin, 2007.

C. Fürweger, M. Kufeld, A. Schlaefer, and C. Drexler. Fiducial-free spinal radiosurgery: Patient motion and targeting accuracy in 227 single fraction treatments with the Cyberknife. *IFMBE Proceedings, World Congress on Medical Physics and Biomedical Engineering, 2009*, 25/1, 277–280, 2009.

T. Guerrero, G. Zhang, W. Segars, et al. Elastic image mapping for 4-D dose estimation in thoracic radiotherapy. *Radiat. Prot. Dosimetry*, 115, 497–502, 2005.

H. W. Hamacher and K.-H. Küfer. Radiation therapy planning—a multicriteria linear programming problem. *Discrete Appl. Math.*, 118, 145–161, 2002.

S. Hatipoglu, Z. Mu, D. Fu, and G. Kuduvalli. Evaluation of a robust fiducial tracking algorithm for image-guided radiosurgery. In *Medical Imaging 2007: Visualization and Image-Guided Procedures*, edited by K. R. Cleary, M. I. Miga, *Proceedings of SPIE*, 6509, 65090A, 2007.

M. Hoogeman, J.B. Prévost, J. Nuyttens, J. Pöll, P. Levendag, B. Heijmen. Clinical accuracy of the respiratory tumor tracking system of the cyberknife: assessment by analysis of log files. *Int. J. Radiat. Oncol. Biol. Phys.*, 74(1), 297–303, 2009.

K. -W. Jee, D. L. McShan, and B. A. Fraass. Lexicographic ordering: intuitive multicriteria optimization for IMRT. *Phys. Med. Biol.*, 52, 1845–1861, 2007.

T. Juhler-Nottrup, S. S. Korreman, A. N. Pedersen, L. R. Aarup, H. Nystrom, M. Olsen, and L. Specht, Intra- and interfraction breathing variations during curative radiotherapy for lung cancer. *Radiother. Oncol.*, 84, 40–48, 2007.

W. A Kalender. X-ray computed tomography. *Phys. Med. Biol.*, 51, R29–R43, 2006.

P. J. Keall, S. Joshi, S. S. Vedam, J. V. Siebers, V. R. Kini, and R. Mohan. Four-dimensional radiotherapy planning for DMLC-based respiratory motion tracking. *Med. Phys.*, 32 (4), 942–951, 2005.

U. W. Langner and P. J. Keall. Prospective displacement and velocity-based cine 4D CT. *Med. Phys.*, 35, 4501, 2008.

J. G. Li and L. Xing. Inverse planning incorporating organ motion. *Med. Phys.*, 27, 1573–1578, 2000.

D. A. Low, M. Nystrom, E. Kalini, P. Parikh, J. F. Dempsey, J. D. Bradley, S. Mutic, S. H. Wahab, T. Islam, G. Christensen, D. G. Politte, and B. R. Whiting. A method for the reconstruction of four-dimensional synchronized CT scans acquired during free breathing. *Med. Phys.*, 30, 1254–1263, 2003.

A. E. Lujan, E. W. Larsen, J. M. Balter, et al. A method for incorporating organ motion due to breathing into 3D dose calculations. *Med. Phys.*, 26, 715–720, 1999.

Y. Ma, L. Lee, O. Keshet, P. Keall, and L. Xing. Four-dimensional inverse treatment planning with inclusion of implanted fiducials in IMRT segmented fields. *Med. Phys.*, 36(6), 2215–21, 2009.

G. S. Mageras, Z. Fuks, S.A. Leibel, C. C. Ling, M. J. Zelefsky, H. M. Kooy, M. van Herk, G. J. Kutcher. Computerized design of target margins for treatment uncertainties in conformal radiotherapy. *Int. J. Radiat. Oncol. Biol. Phys.*, 43(2), 437–45, 1999.

J. R. McClelland, J. M. Blackall, S. Tarte, A. C. Chandler, S. Hughes, S. Ahmad, D. B. Landau, D. J. Hawkes. A continuous 4D motion model from multiple respiratory cycles for use in lung radiotherapy. *Med. Phys.*, 33(9), 3348–58, 2006.

C. Men, X. Gu, D. Choi, A. Majumdar, Z. Zheng, K. Mueller, S. B. Jiang. GPU-based ultrafast IMRT plan optimization. *Phys. Med. Biol.*, 54(21), 6565–73, 2009.

A. Y. Minn, D. Schellenberg, P. Maxim, Y. Suh, S. McKenna, B. Cox, S. Dieterich, L. Xing, E. Graves, K. A. Goodman, D. Chang, A. C. Koong. Pancreatic tumor motion on a single planning 4D-CT does not correlate with intrafraction tumor motion during treatment. *Am. J. Clin. Oncol.*, 32(4), 364–68, 2009.

A. Muacevic, C. Drexler, B. Wowra, A. Schweikard, A. Schlaefer, R. T. Hoffmann, R. Wilkowski, H. Winter, M. Reiser. Technical description, phantom accuracy, and clinical feasibility for single-session lung radiosurgery using robotic image-guided real-time respiratory tumor tracking. *Technol. Cancer Res. Treat.*, 6(4), 321–28, 2007.

M. J. Murphy and S. Dieterich. Comparative performance of linear and nonlinear neural networks to predict irregular breathing. *Phys. Med. Biol.*, 51(22), 5903–14, 2006.

T. Pan, T. Y. Lee, E. Rietzel, et al. 4D-CT imaging of a volume influenced by respiratory motion on multi-slice CT. *Med. Phys.*, 31, 333–340, 2004.

C. Plathow, S. Ley, C. Fink, M. Puderbach, W. Hosch, A. Schmahl, J. Debus, and H. U. Kauczor. Analysis of intrathoracic tumor mobility during whole breathing cycle by dynamic MRI. *Int. J. Radiat. Oncol. Biol. Phys.*, 59(4), 952–959, 2004.

D. Putra, O. C. Haas, J. A. Mills, K. J. Burnham. A multiple model approach to respiratory motion prediction for real-time IGRT. *Phys. Med. Biol.*, 53(6), 1651–63, 2008.

H.E. Romeijn, J.F. Dempsey, and J.G. Li. A unifying framework for multi-criteria fluence map optimization models. *Phys. Med. Biol.*, 49(10), 1991–2013, 2004.

A. Schlaefer, J. Fisseler, S. Dieterich, H. Shiomi, K. Cleary, and A. Schweikard. Feasibility of four-dimensional conformal planning for robotic radiosurgery. *Med. Phys.*, 32(12), 3786–92, 2005.

A. Schlaefer, A. Schweikard. Stepwise multi-criteria optimization for robotic radiosurgery. *Med. Phys.*, 35(5), 2094–103, 2008.

E. Schreibmann, M. Lahanas, L. Xing, and D. Baltas. Multiobjective evolutionary optimization of the number of beams, their orientations and weights for intensity-modulated radiation therapy. *Phys. Med. Biol.*, 49(5), 747–70, 2004.

E. Schreibmann, G. T. Y. Chen, and L. Xing. Image interpolation in 4D CT using a B-Spline deformable registration model. *Int. J. Radiat. Oncol. Biol. Phys.*, 64, 1537–50, 2006.

E. Schreibmann, B. Thorndyke, T. Li, J. Wang, L. Xing. Four-dimensional image registration for image-guided radiotherapy. *Int. J. Radiat. Oncol. Biol. Phys.*, 71(2), 578–86, 2008.

Y. Seppenwoolde, R. I. Berbeco, S. Nishioka, H. Shirato, and B. Heijmen. Accuracy of tumor motion compensation algorithm from a robotic respiratory tracking system: A simulation study. *Med. Phys.*, 34, 2774, 2007.

G. C. Sharp, S. B. Jiang, S. Shimizu, and H. Shirato. Prediction of respiratory tumor motion for real-time image-guided radiotherapy. *Phys. Med. Biol.*, 49(3), 425–440, 2004.

Shirato H, et al. Four-dimensional treatment planning and fluoroscopic real-time tumor tracking radiotherapy for moving tumour. *Int. J. Radiat. Oncol. Biol. Phys.*, 48, 435–42, 2000.

J. Sonke, L. Zijp, P. Remeijer, and M. Van Herk. Respiratory correlated cone beam CT. *Med. Phys.*, 32(4), 1176–1186, 2005.

J. Unkelbach and U. Oelfke. Inclusion of organ movements in IMRT treatment planning via inverse planning based on probability distributions. *Phys. Med. Biol.*, 49, 4005–4029, 2004.

J. Van de Steene, N. Linthout, J. de Mey, V. Vinh-Hung, C. Claassens, M. Noppen, A. Bel, and G. Storme. Definition of gross tumor volume in lung cancer: Inter-observer variability. *Radiother. Oncol.*, 62(1), 37–49, 2002.

M.van Herk. Errors and margins in radiotherapy. *Seminars in Radiation Oncology*, Vol. 14, issue 1, pp. 52–64, 2004.

S. S. Vedam, P. J. Keall, V. R. Kini, H. Mostafavi, H. P. Shukla, and R. Mohan. Acquiring a four-dimensional computed tomography dataset using an external respiratory signal. *Phys. Med. Biol.*, 48(1), 45–62, 2003.

M. von Siebenthal, G. Székely, U. Gamper, P. Boesiger, A. Lomax, P. Cattin. 4D MR imaging of respiratory organ motion and its variability. *Phys. Med. Biol.*, 52(6), 1547–64, 2007.

S. Webb: Optimization of conformal radiotherapy dose distributions by simulated annealing. *Phys. Med. Biol.*, 34, 1349–70, 1989.

J. J. Wilkens, J. R. Alaly, K. Zakarian, W. L. Thorstad, and J. O. Deasy. IMRT treatment planning based on prioritizing prescription goals. *Phys. Med. Biol.*, 52:1675–1692, 2007.

Y. Xie, D. Djajaputra, C. R. King, S. Hossain, L. Ma, L. Xing. Intrafractional motion of the prostate during hypofractionated radiotherapy. *Int. J. Radiat. Oncol. Biol. Phys.*, 72(1), 236–46, 2008.

D. Yan, D. A. Jaffray, and J. W. Wong. A model to accumulate fractionated dose in a deforming organ. *Int. J. Radiat. Oncol. Biol. Phys.* 44, 665–75, 1999.

D. Yang, W. Lu, D. A. Low, J. O. Deasy, A. J. Hope, and I. El Naqa. 4D CT motion estimation using deformable image registration and 5D respiratory motion modeling. *Med. Phys.*, 35, 4577, 2008.

T. Zhang, R. Jeraj, H. Keller, W. Lu, G. H. Olivera, T. R. McNutt, T. R. Mackie, and B. Paliwal. Treatment plan optimization incorporating respiratory motion. *Med. Phys.*, 31, 1576–1586, 2004.

Treatment Planning for Motion Management via DMLC Tracking

9.1 Introduction 77
    Dynamic Multileaf Collimator Intensity-Modulated Delivery
9.2 DMLC Tracking Leaf-Sequencing Evolution 78
    Basic Governing Equations • Summary of Algorithm Properties
9.3 DMLC Control Algorithms for 1D Moving and Deforming Targets 79
    Optimal Solutions Developed for a Rigid Moving Target Based on Data Collected Before
    Treatment • Optimal Solutions Developed for a Moving, Deforming Target with Motion Data
    Based on Prior Measurements • Self-Corrected Delivery: Dose Delivery and Motion Model
    Errors
9.4 DMLC Control Algorithms for 3D Moving Targets 81
    Synchronized MLC for Tongue and Groove • Synchronized MLC for Targets Moving in
    3D • Real-Time Synchronized MLC for Targets Moving in 3D
9.5 Toward Motion-Optimized IMRT 84
    Organs at Risk Sparing: DMLC Control Algorithms with Constant Linear Accelerator Dose
    Rate • Organs at Risk Sparing: DMLC Control Algorithms with Variable Linear Accelerator
    Dose Rate
9.6 Motion Management in VMAT 86
9.7 Summary 88
    Acknowledgment 88
    Appendix 9.A Derivation of the Basic Equations for MLC Tracking of Moving Targets 88
    Appendix 9.B Interdependence of Delivery Parameters of VMAT 90
    References 90

Lech Papiez

Dharanipathy Rangaraj

9.1 Introduction

The issues related to organ motion and deformation management in radiation therapy have been a problem during the entire development of the field (George et al. 2003; Huang et al. 2002; Keall et al. 2006; Sawant et al. 2008; Srivastava et al. 2007; Webb 2006; McQuaid and Webb 2006). These issues and the clinical challenges they create are discussed in detail in many chapters of this book. There are several techniques to remove or eliminate effects of tissue motion degradation of the quality of radiation therapy, such as increasing the margin, applying gating, use of robotic tracking, and scanning the beam so that it follows the moving target. In this chapter, we focus on Dynamic Multileaf Collimator (DMLC)-based tracking that aims at improving the Intensity Modulated Radiation Therapy (IMRT) delivery accuracy to moving and deforming patient anatomy. The chapter mainly focuses on delivery algorithms evolving from simple to fairly advanced and complex algorithms. Several novel concepts such as simultaneous tumor

and organs at risk tracking through DMLC leaf sequencing and recent results of DMLC tracking in volumetric modulated arc therapy (VMAT) are described. Finally, thoughts are presented on true 4D planning, and the feasibility of 4D planning is discussed.

9.1.1 Dynamic Multileaf Collimator Intensity-Modulated Delivery

In DMLC IMRT, modulated radiation fluence is delivered in a continuous fashion using multileaf collimators. The goal of the leaf-sequencing algorithm is to determine the appropriate leaf speed as a function of time to deliver the modulated fluence accurately to the intended anatomy.

9.1.1.1 IMRT Delivery to Moving Targets

When the tumor is moving, as is often the case for thoracic and abdominal tumors, the delivery of the intended modulated fluence gets compromised. Researchers have argued that these effects average out over a span of several fractions. But there are also arguments that daily dose accuracy is important for tumor control. In case of hypofractionation, this becomes even more important. Thus, appropriate modification of DMLC leaf sequencing to take into account the moving anatomy to deliver the intended fluence without violating any mechanical constraints is compulsory to improve the quality of IMRT radiotherapy (McMahon et al. 2007, 2008; Papiez and Abolfath 2005; Papiez et al. 2007; Papiez and Rangaraj 2005; Rangaraj et al. 2008; Rangaraj and Papiez 2005; Sun et al. 2010; Papiez 2003, 2004).

9.1.1.2 Why DMLC as a Tracking Solution?

The dynamic intensity-modulated radiation therapy (DMLC IMRT) envisions delivery of modulated irradiation fields to moving human body anatomy in a continuous way, without beam interruption. This type of delivery, if sufficiently accurate, leads to vast improvement in treatment effectiveness compared with gating IMRT applied in the presence of tissue motion. Moreover, while gated IMRT delivers treatment assuming immobility of body anatomy at particular intervals of the periodic motion cycle, in actuality, the lingering motion of anatomy at treatment intervals persists, *which leads to inevitable residual error in dose delivered relative to dose as designed by a plan.* Being aware of the potential advantage of DMLC IMRT, one is prompted to ask the question how the goal of DMLC IMRT delivery may be achieved. The question is considerably simplified if we restrict our attention to dose delivery only to the target, with the exclusion of other tissues exposed. Effects of interaction between intensity modulation and relative motions between the target and other tissues in the body clearly influence the quality of treatment, but they may not be essential for the tumor eradication. In the case when delivery to the target is the preliminary concern, the most natural idea for the imposition of dose to this structure in IMRT comes from the observation that motion of leaf apertures over a static target, if preserved in the target frame of reference when the target is moving, should result in the imposition of the same intensity map over the moving target as it has done over the static target (Keall et al. 2006). However, this approach, though intuitively appealing, is relatively inflexible. First, if the beam dose rate is uneven and the motion of the leaves is enslaved to the beam dose rate rather than time, then keeping a perfect correlation of leaf positions with respect to a moving target according to the pattern derived for a static target will not deliver the static target intensity map to the moving target (Abolfath and Papiez 2009). Moreover, if delivery does not proceed as planned (due to hardware failure or unexpected target motion) the delivery based on the assumption that leaf motion has a unique relation to target position will introduce an uncorrectable error. In actuality, the errors introduced during delivery can be largely compensated if more flexible delivery strategies are adopted (MacMahon et al. 2007, 2008; Rangaraj et al. 2008). Finally, a DMLC strategy based on delivering leaf aperture motion derived for a static target to a moving target can compromise the effectiveness of delivery procedure (Rangaraj and Papiez 2005).

Being aware of these shortcomings of DMLC IMRT to moving targets based on preserving leaf apertures in the target frame of reference, one would like to search for more a generic approach that provides more flexibility in realizing proper IMRT delivery to moving anatomy under variable and unpredictable conditions. This chapter is devoted to deriving these solutions and analyzing their properties and applications.

9.1.1.3 Complications and Issues with DMLC IMRT Delivery to Moving Target

There are multiple complications that must be addressed when one adopts DMLC-based target tracking algorithms:

1. Tumors move in an irregular fashion.
2. Some tumors deform during motion.
3. Not only the tumor but also the organs at risk move.
4. Tumors behave differently from day to day.
5. Linear accelerator is a pulsed beam, and the ability to maintain constant dose rate would be compromised during the delivery, which introduces delivery errors.
6. Today's MLC design is such that the leaves move only in one direction while tumors move in 3D [in 2D in beam's eye view (BEV) plane of the fixed angle beam, which imposes variable beam fluence on the target from the given beam direction].
7. Because of the dynamic nature of the DMLC tracking delivery and the real-time motion monitoring system, unintended errors might occur during delivery and be propagated or enhanced as the delivery progresses.
8. The MLC delivery system, linac feed back system, and real-time motion monitoring system all have some latency associated with it.
9. Delivery system (MLC) has physical limitations such as leaf speed compared with the nature of the tumor motion speed, deformation, irregularity, etc.
10. Gantry has physical limitations in gantry speed and acceleration in VMAT delivery.
11. Delivery time increases for tracking deliveries.

So the question is how to address all these issues that are associated with DMLC tracking delivery. While some of these are general issues for any tracking delivery solution, some others are unique to DMLC delivery. In the following solution, we have described delivery ideas, concepts, and solutions to address several of the above-stated problems using DMLC leaf sequencing to adapt to moving anatomy.

9.2 DMLC Tracking Leaf-Sequencing Evolution

9.2.1 Basic Governing Equations

To keep the flow of presentation of basic ideas uninterrupted, we move technical aspects of the derivation of equations governing motions of MLC for moving target deliveries to Appendix 9.A. The

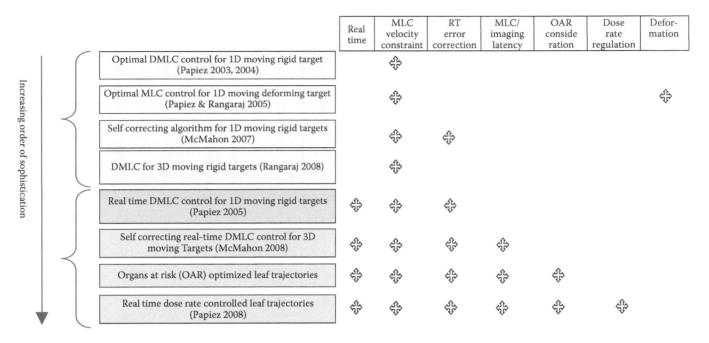

	Real time	MLC velocity constraint	RT error correction	MLC/ imaging latency	OAR conside ration	Dose rate regulation	Defor- mation
Optimal DMLC control for 1D moving rigid target (Papiez 2003, 2004)		✜					
Optimal MLC control for 1D moving deforming target (Papiez & Rangaraj 2005)		✜					✜
Self correcting algorithm for 1D moving rigid targets (McMahon 2007)		✜	✜				
DMLC for 3D moving rigid targets (Rangaraj 2008)		✜					
Real time DMLC control for 1D moving rigid targets (Papiez 2005)	✜	✜	✜				
Self correcting real-time DMLC control for 3D moving Targets (McMahon 2008)	✜	✜	✜	✜			
Organs at risk (OAR) optimized leaf trajectories	✜	✜	✜	✜	✜		
Real time dose rate controlled leaf trajectories (Papiez 2008)	✜	✜	✜	✜	✜	✜	

Increasing order of sophistication

FIGURE 9.1 The evolution of the DMLC control algorithms developed by several groups.

readers interested in these derivations before learning about their applications for delivery strategies should turn to Appendix 9.A.

9.2.2 Summary of Algorithm Properties

Figure 9.1 shows the evolution of the DMLC control algorithms developed by several groups. The next sections below highlight the capabilities of the algorithms.

9.3 DMLC Control Algorithms for 1D Moving and Deforming Targets

The algorithms that fall under these categories are for targets that move and deform in one dimension.

9.3.1 Optimal Solutions Developed for a Rigid Moving Target Based on Data Collected Before Treatment

Assumptions: Target moving in 1D; no deformation; motion is known *a priori*; and constant dose rate.

Constraints: Deliver predefined Intensity profile; do not exceed MLC maximum leaf velocity.

The initial algorithms we developed for the optimization of DMLC IMRT delivery assumed rigidly translating target volumes (Papiez 2003, 2004). In these articles, interdependent differential equations of motion for MLC leaf pairs were developed in order to minimize beam-on time for DMLC IMRT delivery assuming rigidly translating target volumes. The derivations assumed continuous, unidirectional MLC leaf motions and constant dose-rate

delivery. The strategies were intended to minimize the number of monitor units for each beam. The resulting solutions reproduced the original intensity profiles for each leaf pair. To solve this problem, the independent leaf motion formulation was devised as a 1D optimization problem with two constraints. One constraint required that the predefined fluence profiles be delivered by individual leaf pairs. The other constraint imposed a maximum allowed speed for each leaf in the linear accelerator frame of reference. Thus this algorithm can be used to optimally deliver the desired fluence for 1D moving rigid target with MLCs velocity constraints whose motion during delivery follows the motion as determined from the model developed during simulation.

9.3.2 Optimal Solutions Developed for a Moving, Deforming Target with Motion Data Based on Prior Measurements

Assumptions: Target moving and deforming in 1D, and motion and deformation are known *a priori*.

Constraints: Deliver predefined Intensity profile; do not exceed MLC maximum leaf velocity.

The leaf-sequencing algorithm was extended to account for periodic target deformation based on measurements acquired during simulation (Papiez and Rangaraj 2005; Papiez et al. 2005). The algorithm was designed to use the periodic deformation as direct input into the leaf-sequencing step in IMRT planning. The model was tested on arbitrary intensity profiles and allowed any type of uniform expansion and contraction of the target during the breathing cycle and relative to a reference shape (e.g., during exhalation). The algorithm output was a replacement of the

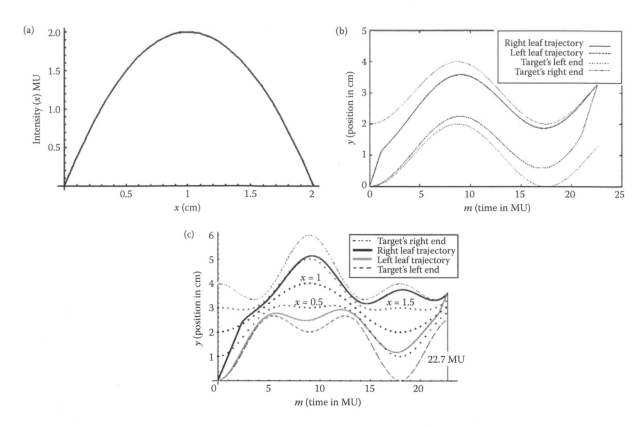

FIGURE 9.2 (a) Intensity profile to be delivered over the target; (b) the leaf trajectories to deliver the intensity shown in (a) if the target experiences a rigid motion known *a priori*; (c) the leaf trajectories to deliver the intensity shown in (a) if the target experiences a moving and oscillatory deformation motion known *a priori*. Note: The parabola is used for illustrative purposes. The algorithm can manage arbitrary intensity and motion profiles. See Papiez (2003, 2004), Papiez and Rangaraj (2005), Papiez et al. (2005), and Appendix 9.A for further details.

original static MLC leaf sequence with a sequence that would be used at the time of treatment. An additional goal of the optimization was to minimize the beam-on time (maximize the treatment duty cycle). *Thus, this algorithm can be used to deliver the desired fluence for 1D moving and deforming targets with MLCs velocity constraints whose motion and deformation during delivery follows as determined from the model developed during simulation.*

Figure 9.2 shows examples of a test fluence profile, leaf motion trajectories for the rigid target motion, and leaf motion trajectories for the moving and deforming target.

9.3.3 Self-Corrected Delivery: Dose Delivery and Motion Model Errors

Assumptions: Target moving in 1D, and motion and deformation are unknown (real-time delivery).

Constraints: Deliver predefined intensity profile; do not exceed MLC maximum leaf velocity.

DMLC treatments are not typically delivered in time exactly as the leaf sequence prescribes. For example, one leaf may not be able to move at its specified velocity; hence, the entire delivery sequence must slow down to accommodate the sticking leaf. Organ motions

inside the body registered before treatment and during treatment are not identical. Observations of tissue motion at the time of treatment have revealed that the motion pattern can change from fraction to fraction. The optimization algorithm was extended to account for delivery errors such as this, as well as the changes in the tumor motion model as measured in real time (Papiez et al. 2005). The solutions were based on the following scheme: (a) an initial solution is found given the motion model obtained from the CT simulation scans and specifications for linear accelerator performance; (b) the solution developed in "a" is restricted to the domain of leaf velocities where each trailing leaf is maintained at a predetermined speed that is less than the maximum allowable leaf speed; (c) the local target velocity (as well as the local deformation of the target) is derived in real time from observation of measured surrogates monitoring the patient respiratory cycle; and (d) the leading and trailing leaf velocities are adjusted by the projected target velocities at the leaf positions. The purpose of the constraint in "a" is to provide some bandwidth for the trailing leaf velocities in case their velocities need to increase. The leading leaf velocities are subsequently limited by adjusting the overall dose rate of the linear accelerator. *Thus, this algorithm allows for the control of MLC leaves based entirely on real-time calculations (instead of the motion model determined during simulation) of the fluence delivered over the target. The algorithm is capable of efficiently*

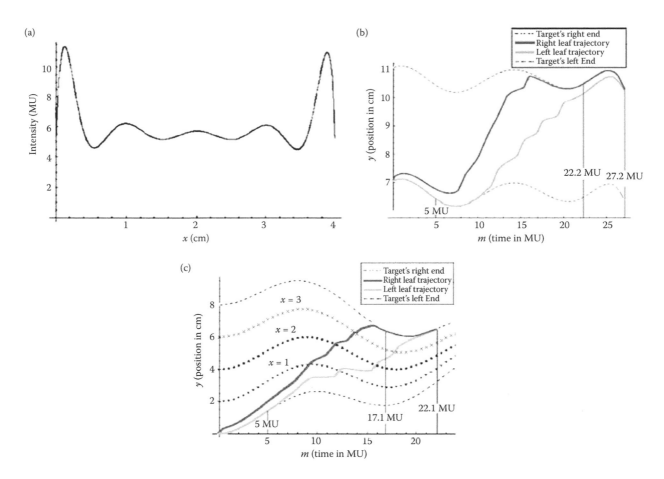

FIGURE 9.3 (a) Fluence (intensity) profile to be delivered over the target by a single leaf pair; (b) the leaf trajectories to deliver the fluence shown in (a) in real time if the target experiences a rigid motion not known *a priori*; (c) the leaf trajectories to deliver the fluence profile shown in (a) in real time if the target experiences both motion and deformation not known *a priori*. Note: The parabola is used for illustrative purposes. The algorithm can manage arbitrary intensity and motion profiles. See Papiez et al. (2005) and Appendix 9.A for further details.

correcting generalized delivery errors without requiring the interruption of delivery where a generalized delivery error (linear accelerator related or target motion related) between the delivered and intended fluence intensity profiles.

Figure 9.3 shows an example of derived fluence profile of a single pair and corresponding leaf motion trajectories.

9.4 DMLC Control Algorithms for 3D Moving Targets

The algorithms that fall under these categories are for targets that move in three dimensions.

9.4.1 Synchronized MLC for Tongue and Groove

Assumptions: Target moving in 1D; no deformation; motion is known *a priori*; and constant beam dose rate.

Constraints: Deliver predefined intensity profile; do not exceed MLC maximum leaf velocity; avoid tongue-and-groove effect for all the leaf pairs.

This algorithm was originally designed to limit the tongue-and-groove effect in DMLC leaf sequences. To remove the tongue-and-groove effect (Rangaraj and Papiez 2005), neighboring leaves were synchronized such that they no longer were allowed arbitrary relative motion patterns. This led to slightly increased monitor units, but did improve the accuracy of dose distribution calculations, which were unable to model the tongue-and-groove effect. The synchronization process also led to MLC sequences with the unintended benefits that are useful in the following: (1) they assured that leading and following leaf ends did not pass by neighboring leaf ends (thus interdigitation is removed); and (2) they lead to radiation portals that tended to have less jagged portal boundaries. *The property of delivery through fields with smoother edges allowed, in turn, the unintended consequence of reducing dose delivery errors caused by projected target motion perpendicular to the direction of MLC leaf motion, which is the reason that the algorithm is mentioned here.* Subsequently, synchronized DMLC leaf-sequencing algorithms have been designed so that they are applicable for general cases of IGIMRT delivery to deforming targets and for real-time target motion data delivery.

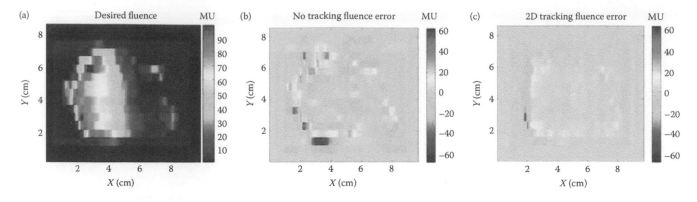

FIGURE 9.4 **(See color insert.)** (a) Desired fluence distribution; (b) fluence error without motion compensation when the target experiences elliptical motion of 2.0 (parallel) × 0.5 cm (perpendicular); (c) fluence error with 3D synchronized motion tracking.

9.4.2 Synchronized MLC for Targets Moving in 3D

Assumptions: Target moving in 3D (2D in the projected beam space); no deformation; motion is known *a priori*; and constant dose rate.

Constraints: Deliver predefined intensity profile; do not exceed MLC maximum leaf velocity, multiple leaf-pair motion dependence.

In this algorithm (Rangaraj et al. 2008), the motion of the target is divided into two components. First, the leaf trajectories are modified considering only the target motion component in the leaf travel direction. Second, the synchronization algorithm described in Section 9.4.1 is applied (Papiez 2003, 2004). Because the MLC leaves move along a single axis, motion compensation along the direction perpendicular to the leaf motion is accomplished by translating the leaf positions from one leaf bank to the next to match the target motion along that axis. The use of the synchronization algorithm serves to reduce the leaf-to-leaf position variations inherent in unsynchronized leaf sequences. *Small dosimetric errors may occur while switching depending on the time it takes to do the translation and the discrimination of the target motion necessitated by the finite leaf width. Dosimetric errors are most visible in the treatment field regions where greatest variations of intensity prevail, as those large variations make difference in leaf positions larger and thus influence the leaf switching efficiency. It is worth to notice that these errors result from limitations of MLC ability to follow the moving target in 2D in BEV.* It has been proved (Rangaraj et al. 2008) that algorithms for 2D tracking eliminate the dosimetric error completely in case when MLC speed constraints vanish and width of leaves are appropriately diminished.

Figure 9.4 shows the fluence distribution from a portal used in a lung cancer treatment plan. The impact of breathing motion on the resulting fluence distribution is shown for an elliptical motion trajectory of 2 cm in the parallel and 0.5 cm in the orthogonal orientations to the leaf motion direction, respectively. The results show that without the correction,

TABLE 9.1 Statistical Assessment of Delivery Errors when None or Limited Compensation is Applied for Target Motion in Radiotherapy Delivery

(a)

Tracking Strategies (MU)	No Tracking	2D Motion Tracking
RMS difference	15.67	5.00
Standard deviation	15.66	4.88

(b)

Maximum Leaf Velocity/ Error Statistics (MU)	1 cm/s	2 cm/s	3 cm/s	4 cm/s	5 cm/s
RMS difference	5.77	5.15	4.95	4.9	4.82
Standard Deviation	5.75	5.1	4.90	4.84	4.80

root-mean-square fluence errors of as large as 16MUs occur over relatively large areas of the radiation portal. Applying the synchronized 3D motion compensation algorithm reduces the errors to less than 6 MU throughout the portal. Table 9.1 shows the error statistics and the influence of the MLC maximum leaf velocity on the fluence error (higher the maximum leaf velocities, lower the fluence error).

Table 9.1a shows the error statistics summary. Table 9.1b shows the impact of varying the maximum MLC leaf velocity on the error statistics.

9.4.3 Real-Time Synchronized MLC for Targets Moving in 3D

Assumptions: Target moving in 2D in beam's eye view; no deformation; motion is unknown before treatment (real-time motion registration).

Constraints: Deliver predefined intensity profile; do not exceed MLC maximum leaf velocity; avoid tongue-and-groove effect for all the leaf pairs.

This MLC control algorithm was developed for delivering IMRT to targets that are undergoing 3D motion (2D in Beam's

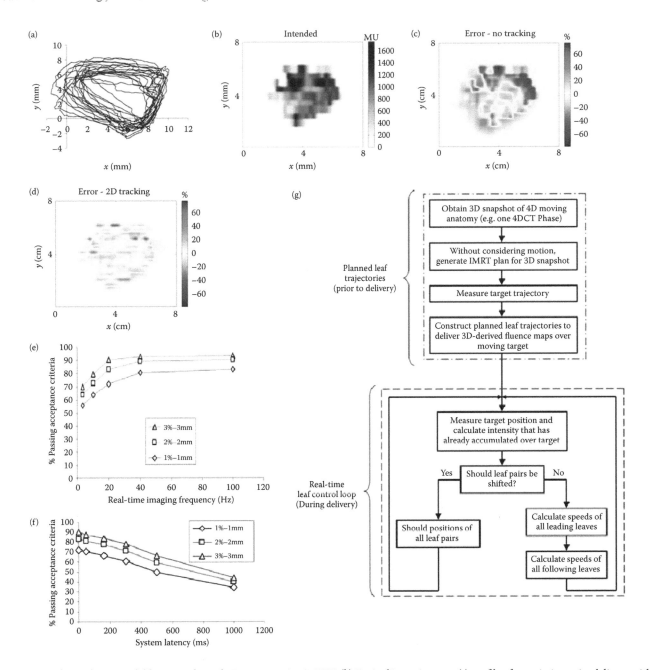

FIGURE 9.5 **(See color insert.)** (a) Patient-derived 2D target motion in BEV; (b) Desired intensity map; (c) profile of error in intensity delivery without tracking; (d) profile of error in intensity delivery with 2D tracking in real time based on feedback; (e) dependence of delivery quality on the frequency of target position measurement; (f) dependence of delivery quality on real-time tracking latency of the system. See McMahon et al. (2008) for further details.

eye view plane) in real time (McMahon et al. 2008). The goal of this method is to deliver 3D-derived fluence maps over a moving patient anatomy. Target motion measured prior to delivery is first used to design a set of planned DMLC leaf trajectories. During actual delivery, the algorithm relies on real-time feedback to compensate for target motion that does not agree with the motion measured during planning. The methodology is based on an existing 1D algorithm that uses on-the-fly intensity calculations to appropriately adjust the DMLC leaf trajectories in real time during exposure delivery (Papiez 2003, 2004). To extend the 1D algorithm's application to 2D target motion, a real-time

leaf-pair shifting mechanism has been developed. Target motion that is orthogonal to leaf travel is tracked by appropriately shifting the positions of all MLC leaves (Rangaraj et al. 2008). *This algorithm is robust in the sense that it does not rely on a high level of agreement between the motion predicted by motion model developed during simulation and the actual motion delivery.*

The performance of the tracking algorithm was tested for a single beam of a fractionated IMRT treatment, using a clinically derived intensity profile and a 2D target trajectory based on measured patient data. Figure 9.5 shows (a) patient derived 2D target motion in BEV, (b) desired intensity map, (c) profile of error in intensity delivery without tracking, and (d) profile of error in intensity delivery with 2D tracking in real time based on feedback. The dependence of delivery quality on the frequency of target position measurement, shown in Figure 9.5e, shows that the higher the frequency, the higher the passing rate. Also the dependence of delivery quality on real-time tracking latency of the system is shown in Figure 9.5f.

9.5 Toward Motion-Optimized IMRT

The algorithms that fall under this category deal with deliveries that take into account not only dose imposed over the target but also dose to the critical structures (Papiez et al. 2007). The delivery process times the MLC leaf positions in their motion with respect to moving targets and differently moving organs at risk and simultaneously varies the beam dose rate so as to achieve a cumulative dose to irradiated volumes. The algorithms result in quality of delivery higher than that of the dose deliverable to static body structures when the factor of timing beam parameters relative to motion of anatomy is absent.

9.5.1 Organs at Risk Sparing: DMLC Control Algorithms with Constant Linear Accelerator Dose Rate

Assumptions: Target moving in 1D; no deformation; motion is known; OAR static; constant beam dose rate; intended intensity map delivered over target; and minimized intensity delivered to OAR.

Constraints: Deliver predefined fluence (intensity) profile to target; do not exceed MLC maximum leaf velocity; multiple leaf pairs trajectories optimized to minimize intensity over OAR.

Delivering the predetermined fluence profile to the target using an MLC allows for a few variable parameters. These parameters can be controlled and optimized (if needed) so that the initial moment of DMLC delivery with respect to the target breathing phase, the speed of MLC leaves, and beam intensity rate can be chosen to minimize the dose delivered to organs at risk. Interestingly, these parameters could be varied in some interdependent manner without compromising the fluence map delivered to the moving target. Thus, there exist multiple solutions to deliver the same fluence map to the moving target, which in turn will change the dose delivered to other organs that are experiencing differential motions. The most interesting among various solutions of this type is the one that assures the minimal dose to organs at risk. Thus, this algorithm (Papiez et al. 2007) defines an MLC leaf sequence with adjustable parameters of speed that allow optimizing (minimizing) dose delivered to sensitive organs while delivering the planned (in 3D treatment planning approach) dose to moving targets. Formulations of optimization procedures based on adjustable parameters of leading leaf speeds that minimize criteria related to the level of exposure of sensitive organs have been investigated for a few simple examples (Papiez et al. 2007). The data clarifying

the quantitative gain that these delivery strategies are able to achieve are collected for various scenarios of treatment objectives and types of organ motions. *All of the previously described algorithms developed by our group (McMahon et al. 2007, 2008; Papiez and Abolfath 2008; Papiez et al. 2007; Papiez and Rangaraj 2008; Rangaraj et al. 2008; Rangaraj and Papiez 2005; Sun et al. 2010; Papiez 2003, 2004) and others have followed the common perception that effects of motion during radiation therapy should be eliminated to achieve the best possible treatment. However, in this algorithm (Papiez et al. 2007), we have shown that it is actually possible to utilize tissue motion for the benefit of the radiation therapy.*

Figure 9.6a shows a snapshot of target (4 cm) and an OAR (1 cm) as visible in the BEV. Figure 9.6b shows the motion of the target relative to the organ at risk and points inside the target and a stationary (in the OAR frame of reference) OAR in the center. In Figure 9.6c, the solid line shows the intensity profile planned over OAR and the dotted line shows the intensity delivered to OAR when simple motion-optimized IMRT delivery is performed, while the black dashed line and gray dashed line lines show the effect of the starting phase of delivery on delivered intensity to the target. Table 9.1 shows in numerical terms that the dose to OAR for a moving target using motion-optimized IMRT delivery is reduced (superior dose distribution) relative to best treatment achievable for the static anatomy delivery. In numerical terms, the integrated intensity delivered to the OAR by motion-optimized IMRT delivery strategy (Figure 9.6c, dotted line) reduced by 42.2% relative to the static delivery (Figure 9.6c, solid line).

9.5.2 Organs at Risk Sparing: DMLC Control Algorithms with Variable Linear Accelerator Dose Rate

Assumptions: Target moving in 1D or 2D; no deformation; motion is known; variable beam dose rate.

Constraints: Deliver predefined intensity profile to the target; do not exceed MLC maximum leaf velocity; dose-rate modulation limitation satisfied; and simultaneous adjustment of multiple leaf-pair velocities.

The next advancement toward motion-optimized IMRT (i.e., discerning dose to target and organs at risk) is to discard the constraint of starting from 3D plan (Papiez and Abolfath 2008). The optimization of IMRT plans that take explicitly into account the motions of body anatomy can provide better solutions compared with 4D delivery strategies based on 3D plans. The explicitly dynamic dose planning and delivery is a truly motion-optimized IMRT. The most difficult part of motion-optimized IMRT is the delivery based on MLC leaves, particularly the timely redistribution of intensity maps over various phases of body anatomy motions. *The factor that decisively helps to make such deliveries possible is the utilization during treatment delivery of the beam dose-rate variation and its proper linking to other parameters of delivery, to assure the*

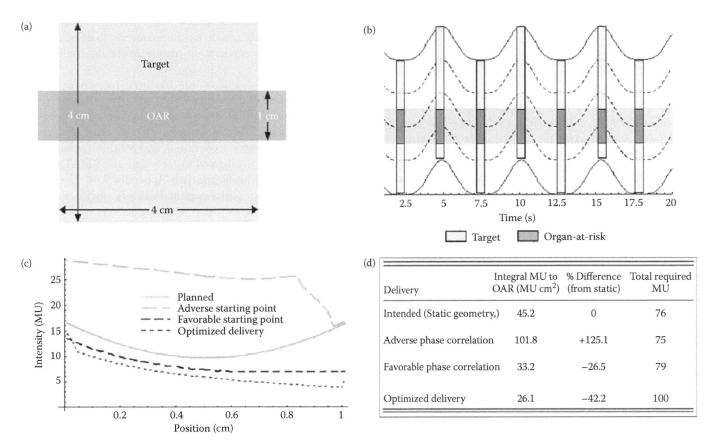

FIGURE 9.6 (a) Snapshot of target (4 cm) and an OAR (1 cm) as visible in BEV; (b) model of the motion of the target and the static OAR (notice that the OAR position (gray band) changes with respect to the moving target); (c) the intensity delivered to the OAR for different cases; (d) comparison of integral MU delivered for OAR. See Papiez et al. (2007) and McMahon et al. (2007) for details.

realization of motion-optimized IMRT plan as well as efficiency of treatment.

To study the deliveries that allow utilization of a variable beam dose rate, we developed an algorithm based on 3D planning as a step away from motion-optimized IMRT. First, we were able to show that our solutions with a time-dependent beam dose-rate delivery lead to DMLC IMRT therapy that increased the accuracy and efficiency of IMRT treatments for stationary and moving targets (Papiez and Abolfath 2008). The method of time-dependent intensity rate DMLC IMRT relies on adapting the leading and following leaf velocities over their subsequent positions over the target with the intensity rate (dose rate) of the linear accelerator. Allowing the leading leaf velocity over subsequent points over the target and beam intensity rate to vary provides two free parameters to modulate the strategy of IMRT map delivery to the target. This method allows choosing the dose delivered to moving healthy tissue that is advantageous for the treatment. It is worth pointing out that those dependencies between leaf speed and dose-rate parameters that we use are dependent on the motion of the target departing critically from relations between speeds of leaves and beam dose rate developed by a Varian DMLC delivery system that assumes immobility of the target. Interestingly, one may find that the Varian delivery methodology based on enslaving leaf velocities to the cumulative MU delivered, rather than

allowing it to be variable in time, while working perfectly for static targets, can introduce extensive errors in DMLC IMRT delivery to moving targets. This ability to choose the leading leaf speed and the beam intensity rate creates a situation when truly 4D optimized therapies can be achieved with DMLC IMRT irradiations. In particular, considerable technical simplification of the described DMLC IMRT delivery problem may be implemented if adjustment of intensity rate is treated independently from other parameters and treated as straightforward scaling of the time axis. This approach separates a two-parameter functional optimization into a simpler problem of two separate one-parameter optimizations. In our earlier preliminary investigations we have studied so far, a one-dimensional problem of variable intensity optimization showing, nevertheless, a considerable potential for improving the efficiency and the quality of IMRT treatments with the approach exploiting variable beam intensity rate. The results of these investigations are summarized in two recent articles, where the intricacies of the method are explained and illustrated with number of examples (Papiez and Abolfath 2008; Abolfath and Papiez 2009). Recently, a new method called dose-rate-regulated tracking has been developed (Yi et al. 2008). This new method applied a preprogrammed dynamic MLC sequence in combination with real-time dose-rate control to simplify the real-time control.

9.6 Motion Management in VMAT

Assumptions: Target moving in 2D in beam's eye view; no deformation; motion is known; variable beam dose rate and gantry speed.

Constraints: Deliver predefined intensity profile to the target; do not exceed MLC maximum leaf velocity, Gantry speed and acceleration limitation satisfied, and dose rate modulation limitation satisfied.

As a new IMRT treatment technique, VMAT has gained widespread interest in the radiation therapy community (Otto 2008; Verbakel et al. 2009; Palma et al. 2008; Tang et al. 2010; Ulrich et al. 2009; Wu et al. 2010; Bortfeld and Webb 2009; Court et al. 2010). VMAT delivery involves continuously rotating gantry and widely varying gantry rotation speed, dose rate, and MLC leaf position. When target motion compensation is incorporated, all these delivery parameters require interactive modulations. To ensure that the treatment plan can be delivered without any change in planned dose distribution for the target and without violating limitations of hardware parameters of a linac, a tracking solution needs to consider a number of machine constraints such as maximal permissible gantry speed, dose rate,

and MLC leaf velocity. In addition, as VMAT often requires the gantry to accelerate and decelerate frequently to deliver a treatment plan, gantry acceleration constraint should also be taken into account.

A DMLC tracking algorithm has been recently developed, which is capable of delivering VMAT to targets that experience 2D motions in the BEV (Sun et al. 2010). The interdependence between several linac parameters, such as dose rate, gantry angular speed, gantry acceleration, and MLC leaf velocity, and target motion is investigated to formulate the VMAT delivery problem for moving targets (see Appendix 9.B). The goal of this algorithm is to achieve optimal delivery efficiency (i.e., the shortest delivery time while minimizing delivery error resulting from target motion).

Logical flow of the DMLC tracking algorithm is shown in Figure 9.7. The tracking along the direction perpendicular to leaf motion was described in Rangaraj et al. (2008). Figure 9.8a–d shows optimal delivery parameters a VMAT plan. Red solid lines represent the maximal permissible values for each parameter. For this VMAT tracking delivery, a delivery time of 95.6 seconds can be achieved, which is only 30.2 seconds longer than that needed for delivery without tracking. For the optimal delivery, that is, the shortest delivery time, at

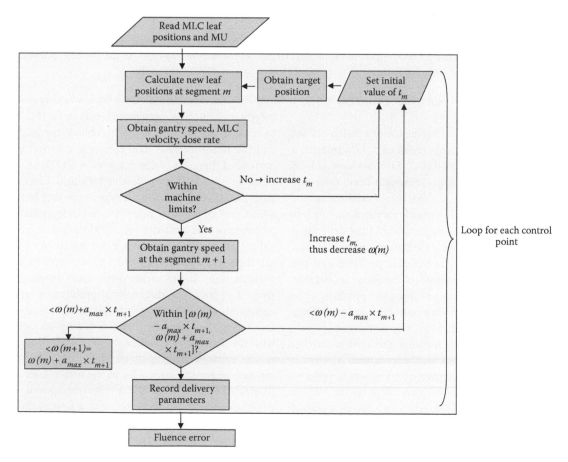

FIGURE 9.7 Logic flow of the DMLC tracking algorithm for VMAT delivery to targets moving in 2D in beam's eye view.

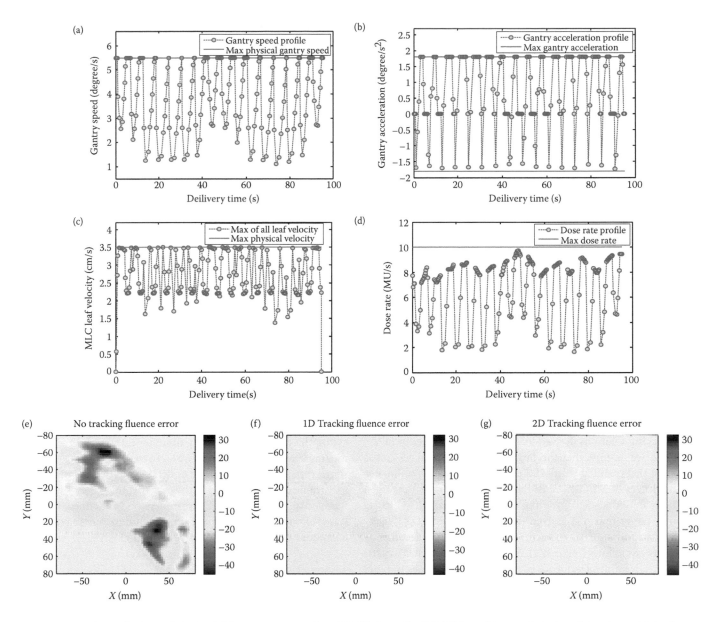

FIGURE 9.8 **(See color insert.)** Optimal delivery parameters for a VMAT plan with 2D DMLC tracking as a function of time in a 360° single arc. (a) Gantry speed, (b) gantry acceleration, (c) MLC leaf velocity, and (d) dose rate. The discrepancies between the planned and delivered fluence for (e) no tracking, (f) 1D tracking, and (g) 2D tracking. See Rangaraj et al. (2008), Sun et al. (2010), and Appendix 9.B for details.

least one of the machine parameters reached its maximum at certain beam angles. The delivery time is mainly determined by the gantry speed, which is strongly modulated so that the MLC leaves can have sufficient time to reach the desired positions. The dose rate and MLC leaf velocity are also significantly modulated as all these delivery parameters are correlated. The discrepancies between the planned and delivered fluence are displayed in Figure 9.8e–g. When the plan is delivered without any tracking, a clear mismatch can be recognized. The fluence is severely blurred along the path of motion. The blurring is significantly removed by employing 1D tracking along the x direction, and best match is achieved with 2D tracking. The

average root-mean-square (RMS) difference of the five plans is 14.8 MU for no motion tracking, 3.6 MU for 1D tracking, and 2.1 MU for 2D motion tracking. *These simulation data indicate that 2D DMLC tracking can achieve high delivery accuracy to a moving target. The algorithm demonstrates that VMAT treatment to a moving target using 2D DMLC motion tracking can be accurately and efficiently delivered with only half a minute increase in delivery time compared to delivery without motion tracking.*

Figure 9.7 shows the logic flow of the DMLC tracking algorithm for VMAT delivery to targets moving in 2D in beam's eye view.

9.7 Summary

The motion management through DMLC tracking has been reviewed in this chapter. Tracking algorithms and some concepts have been discussed from 1D to 3D, from rigid body to deforming target, and from conventional IMRT to VMAT. The DMLC leaf sequencing for the target motion are presented for the target motion in real time and when *a priori* motion information is known. Optimal solutions in terms of minimal treatment time and sparing of OARs are also provided. It has been demonstrated that DMLC is an attractive solution for motion management in radiation therapy.

Acknowledgment

The authors would like to thank Dr. Baozhu Sun for his help in preparing this contribution.

Appendix 9.A Derivation of the Basic Equations for MLC Tracking of Moving Targets

It is instructive first to discuss delivery when target (and surrounding tissues) moves in BEV in one direction. There are numerous situations in which motion is predominantly along one direction in BEV, and so our solutions derived for 1D case target motions may be directly applicable for DMLC IMRT, without invoking solutions derived for 2D motion of tissues in BEV.

When we assume that the target moves in 1D in BEV, it is apparent that tracking with MLC requires turning the collimator so that leaves of MLC move in the same direction as the target. First, we note that delivering the 2D intensity map $I(x,y)$ to 1D moving target makes it convenient to decompose 2D intensity map into a union of 1D intensity maps $I_{yi}(x,y)$. This decomposition works as follows (Figure 9.1). In the planar domain D of variables (x,y), we draw a sequence of parallel to x-axis lines y_i. Lines y_i define bands inside domain D that contain points between lines $y(i)$ and $y(i + 1)$. We denote these bands appropriately as B_{yi}; that is, $B_{yi} = \{(x,y)$ in D; $y(i) < y < y(i + 1)\}$. We assume that distance between any lines $y(i)$ and $y(i + 1)$ $(i = 1,2, k)$ is same as the width of MLC leaves. Let us define now a sequence of functions $I_{yi}(x)$ defined in each band B_{yi} as $I_{yi}(x) = 0.5[I(x,y(i)) + I(x,y(i + 1))$. The collection of functions $I_{yi}(x)$ defined over all B_{yi} is therefore closely approximating original intensity function $I(x,y)$ over its full domain D. We notice here that planning system defines function $I(x,y)$ typically not as a continuous function of x and y, but as a step function. Thus, if the size of the grid in y direction for determining the intensity function $I(x,y)$ from the planning system is chosen to be equal to the width of the lMLC leaf, then the collection of functions $I_{yi}(x)$ defined over all B_{yi} is actually not an approximation of the original intensity function $I(x,y)$, but its exact representation.

The condition of imposing given intensity function $I_{yi}(x)$ along a band B_{yi} by a pair of leaves $(L,F)_{yi}$ is equivalent to keeping the distance of monitor units along axis of intensity between curves $x_L(m)$ and $x_F(m)$ equal exactly to $I_{yi}(x)$ at each x. In other words, if $m_L(x)$ is a cumulative number of MU at which leaf L

passes over x and $m_F(x)$ is a cumulative number of MU at which leaf F passes over x (where by $m_L(x)$, we denote the inverse function to monotonic trajectory $x_L(m)$ and by $m_F(x)$ we denote the inverse function to monotonic trajectory $x_F(m)$), then we have to satisfy the condition

$$m_F(x) - m_L(x) = I_{yi}(x) \qquad (9.A.1)$$

to assure the delivery of intensity $I_{yi}(x)$ defined by leaf trajectories $x_L(m)$ and $x_F(m)$ for pair of leaves $(L,F)_{yi}$.

Differentiating equality (9.A.1) with respect to x, and remembering that $m_L(x)$ and $m_F(x)$ are inverse to $x_L(m)$ and $x_F(m)$, we find that

$$\frac{d}{dm}x_L(m)\big|_x = \frac{(d/dm)x_F(m)\big|_x}{1 - (d/dx)I(x)\cdot(d/dm)x_F(m)\big|_x}$$

or

$$v_L(x) = \frac{v_F(x)}{1 - (d/dx)I(x)\cdot v_F(x)}, \qquad (9.A.2)$$

where $(d/dm)x_F(m)\big|_x = v_F(x)$ and $(d/dm)x_L(m)\big|_x = v_F(x)$ are speeds of following and leading leaves when their trajectories pass point x on the target $(x \in B_{yi})$. Equation 9.A.2 is equivalent to Equation 9.A.1, and we note that it has to be satisfied for each $(x \in B_{yi})$ to assure the delivery of intensity function $I_{yi}(x)$ over $(x \in B_{yi})$ by moving leaves L and F of pair $(I,F)_{yi}$. In the case that the target is static, the speeds in (9.A.2) are understood as defined in laboratory frame of reference and in the case that the target is moving, the speeds in Equation 9.A.2 are understood as defined in target frame of reference. Rewriting formula 9.A.2 for moving target in laboratory frame of reference gives Equation 9.A.3:

$$v_F(x) - v_T^F(x) = \frac{v_L(x) - v_T^L(x)}{1 + \frac{d}{dx}I(x)\cdot[v_L(x) - v_T^L(x)]}, \qquad (9.A.3)$$

where $v_F(x)$ is the speed of the following leaf in laboratory system of coordinates when it passes over point $(x \in B_{yi})$, $v_L(x)$ is the speed of the leading leaf in laboratory system of coordinates when it passes over point $(x \in B_{yi})$ and $v_T^L(x)$ is the speed of the target when leading leaf passes over point $(x \in B_{yi})$ and finally $v_T^F(x)$ is the speed of the target when following leaf passes over point $(x \in B_{yi})$. What formula (9.A.3) tells us is that it is not necessary to have uniquely defined trajectories $x_L(m)$ and $x_F(m)$ for leading and following leaf motions to impose intensity $I_y(x)$ by pair of leaves $(L, F)_{yi}$ over the moving target, but that it is sufficient for $x_L(m)$ and $x_F(m)$ to have speeds at each x related by Equation 9.A.3. Clearly, parameters dI(x)/dx, $v_T^L(x)$ and $v_T^L(x)$ are given as outside input in formula (9.A.3) (first by plan and two other by motion of the target) while two other parameters $v_T^L(x)$ and $v_T^F(x)$ are two changeable variables constrained only by one requirement: that of satisfying (9.A.3). As parameters $v_T^L(x)$ and $v_T^F(x)$ in (9.A.3) are two while formula (9.A.3) relating one to the other is only one we have a large family of solutions for pairs $(v_T^L(x), v_T^F(x))$ that will match each other in formula (9.A.3). Therefore, the multiplicity of solutions admissible by (9.A.3) provides flexibility of choosing those among them that have specific advantages for IMRT treatments. It is generally beneficial for the therapy to look for the best among these solutions relative to specific aspects of therapy that may be clinically advantageous. However, before we move to these investigations, let us discuss first the generalization of solutions admissible by formula (9.A.3) that are also valid for nonrigid target motions.

Thus, let us assume now a more realistic behavior of the target in its motion. Instead of postulating that the target preserves rigidity in its motion, we allow now the deformation of the target during its motion. Surprisingly, allowing for target deformation during DMLC delivery preserves the flexibility of delivery that is obtained for rigid moving targets. However, deriving the formulas for the case of a deforming target cannot proceed as in Equations 9.A.1 through 9.A.3. The reason is that the rigidity condition was essential to write the relation (9.A.1) and then follow with consequences of this expression in (9.A.2) and (9.A.3). In the general case, when point x not only moves with the whole target but also moves relative to other points in the target (as we allow the target to expand or contract), one has to include these changes when defining the exposure of point x (which is determined by the positions of leaves during their time in the space between the point of interest in the target and the radiation source). To this end, we have to introduce first the internal coordinate system for target points along a band B_{yi}. To properly express the positions of points in the moving and deforming target in the immobile system of reference, one must first index the points of the target in the laboratory system of coordinates. To this end, we introduce notation $y(x,m)$ for the coordinate of the point in laboratory frame of reference that is indexed by coordinate x in the target frame of reference at instant of time equivalent to delivery of cumulative number of monitor units equal to m. Having this notation, it is easy to note (see Figure 9.2a) that the opening of point P indexed in target frame of reference by x at instant m to beam exposure due to leading leaf edge passing over this point can be expressed as

$$y_L(m) = y_T(x,m), \qquad (9.A.4a)$$

where $y_L(m)$ is the laboratory frame of reference coordinate of left edge of leading leaf (moving to right) at instant m. Similarly, at instant $m + \Delta m$ one can note (see Figure 9.2b) that the opening of point P1 indexed in target frame of reference by $x + \Delta x$ at instant $m + \Delta m$ to beam exposure due to leading leaf edge passing over this point can be expressed as

$$y_L(m + \Delta m) = y_T(x + \Delta x, m + \Delta m), \qquad (9.A.4b)$$

where $y_L(m + \Delta m)$ is the laboratory frame of reference coordinate of left edge of leading leaf (moving to right) at instant $m + \Delta m$. Then the screening of point P at instant $m + I(x)$ can be expressed as a coincidence of this point at $m + I(x)$ with the right edge of

the following leaf (moving to right) at $m + I(x)$ (see Figure 9.2c). This condition can be written as

$$y_F(m + I(x) + \Delta m) = y_T(x, m + I(x)), \qquad (9.A.4c)$$

where $y_F(m + I(x) + \Delta m)$ is the laboratory frame of reference coordinate of right edge of the following leaf (moving to right) at instant $m + I(x) + \Delta m$. Again at instant $m + I(x) + \Delta m_1$, one can note (see Figure 9.2d) that the screening of point P1 indexed in target frame of reference by $x + \Delta x$ at instant $m + I(x) + \Delta m_1$ from beam exposure due to following leaf edge passing over this point can be expressed as

$$y_F(m + I(x) + \Delta m_1) = y_T(x + \Delta x, m + I(x) + \Delta m_1), \qquad (9.A.4d)$$

where $y_F(m + I(x) + \Delta m_1)$ is the laboratory frame of reference coordinate of right edge of the following leaf (moving to right) at instant $m + I(x) + \Delta m_1$.

Expanding formulas (9.A.4a), (9.A.4b), (9.A.4c), and (9.A.4d) into Taylor series, we obtain after some algebra relationships between speed of moving target v_T and following and leading leaves speeds v_F and v_L in laboratory frame of reference for any x and any m. They are

$$v_L(m) - v_T^L(x) = \frac{d_T(m,x)}{d_T(x)}\left[v_F(x) - v_T^L(x)\right]$$
$$+ \frac{\{dI(x)/dx\}\{v_L(x) - \{\partial y_T(m,x)/\partial m\}}{d_T(x)}$$
$$\times \left[v_F(x) - v_T^L(x)\right] \qquad (9.A.5)$$

for targets deforming during their motions (with $d_T(m,x)$ describing the speed of translation of target vicinity at x at instant m and $\partial y_T(m,x)/\partial m$ describing the rate of target local deformation at x at instant m. Specifying formula (9.A.5) for the case of vanishing deformation (i.e., transforming the formula for the situation when deformation vanishes), we find that (9.A.5) simplifies to Equation 9.A.6:

$$v_F(x) - v_T^F(x) = \frac{v_L(x) - v_T^F(x)}{1 + \dfrac{d}{dx}I(x)\cdot[v_L(x) - v_T^F(x)]}. \qquad (9.A.6)$$

This is identical (as it should be) to Equation 9.A.3 derived earlier for targets moving in strictly rigid fashion. At this point, we have all equations needed to define strategies of DMLC delivery of a given treatment plan to moving targets (rigid or deforming) during their spatial evolution inside the patient body anatomy, when motion of the target is known. We should stress, however, that above equations defining DMLC delivery strategies have somewhat unintuitive characteristics. Namely, motions of leaves of MLC assembly are understood here as being controlled relative to cumulative number of monitor units delivered and not as functions of time. This makes perfect sense if one realizes that m

and $I(x)$ both measure the cumulative fluence and not time. One may also easily understand why the control of leaf motions in practical delivery systems is enslaved to m rather than to time t. However, this dependence is not completely natural for delivery to moving targets. The moving targets are not evolving naturally relative to cumulative number of monitor units but relative to time variable. Thus, a consistent approach for delivery to moving targets requires first expressing target motion as function of m rather than function of t, or alternatively, expressing motions of leaves as functions of time t rather than m.

Let us notice that if the beam dose rate during delivery is constant, then m and t differ only by a constant parameter and so the relation between Equation 9.A.6 and its time equivalent is only a meter of uniform scaling of variable m in these equations by a constant. On the other hand, if the beam dose rate is variable during delivery, the relation between Equation 9.A.6 and its time equivalent is more complicated. In this case, equations equivalent to rigid target evolution (Equation 9.A.6) can be written as follows (Papiez and Abolfath 2008; Abolfath and Papiez 2009):

$$\frac{v_F(m)}{\alpha(t_{F(x)})} - \frac{v_T(x)}{\alpha(t_{F(x)})} = \frac{v_L(x)/\alpha(t_{L(x)}) - v_T^L(x)/\alpha(t_{L(x)})}{1 + \frac{d}{dx}I(x)\cdot[v_L(x)/\alpha(t_{F(x)}) - v_T^L(x)/\alpha(t_{L(x)})]}.$$

(9.A.7)

Equations 9.A.6 and 9.A.7 constitute a system of fundamental equations for DMLC IMRT delivery to moving body anatomy when target moves in 1D along the axis of leaf motion. They can be adapted to deliveries when target is moving in 2D in BEV as we shall see below. The solutions of these equations provide leaf trajectories that define strategies of IMRT delivery to moving bodies. These solutions are obtained by direct integration of above differential equations (Papiez and Rangaraj 2005). In this review article, we rather concentrate our attention on properties of these solutions and their relevance to accuracy and clinical relevance in IMRT therapy to moving patient anatomy.

Appendix 9.B   Interdependence of Delivery Parameters of VMAT

A VMAT plan is uniquely defined by the shaped apertures A at each control point associated with gantry angle g, denoted by $A(g)$, and by the beam weight prescribing the number of MUs associated with each aperture or segment, denoted by $M(g)$. $A(g)$ is a sequence of $2K$ numbers, $X_{L,k}(g)$ and $X_{R,k}(g)$, which specify the left and right leaf positions of the leaf pairs $k = 1, 2, \ldots, K$ at a given gantry angle g. During the delivery of a VMAT plan, the gantry rotates around the patient while both beam aperture $A(g)$ and beam weight $M(g)$ change as a function of gantry angle g. The relationships between the delivery parameters in VMAT are described in the following:

1) The beam weight $M(g)$ at a given gantry angle g during the delivery can be expressed in terms of gantry speed dg/dt and dose rate dm/dt as:

$$M(g) = \frac{dm}{dg}(g) = \frac{dm}{dt}(g) \cdot \frac{dt}{dg}(g),$$

(9.B.1)

where $m(g)$ is the accumulative beam weight delivered until gantry angle g. Thus the dose rate dm/dt at g is given by

$$\frac{dm}{dt}(g) = M(g) \cdot \frac{dg}{dt}(g),$$

(9.B.2)

This shows that the dose rate dm/dt is related to the gantry angular speed dg/dt at gantry angle g and the $M(g)$ prescribed by the VMAT plan.

2) The change of beam aperture $A(g)$ as a function of gantry angle, $V_k(g) = dX_{L(R),k}(g)/dg$, can be related with gantry speed dg/dt and MLC physical leaf velocity $V_{L(R),k}^{phys}(g)$ as follows:

$$V_k(g) = \frac{dX_{L(R),k}(g)}{dg} = \frac{dX_{L(R),k}(g)}{dt} \cdot \frac{dt}{dg} = \frac{V_{L(R),k}^{phys}(g)}{[dg/dt]}$$

(9.B.3a)

evoking

$$V_{L(R),k}^{phys}(g) = V_k(g) \cdot \frac{dg}{dt}.$$

(9.B.3b)

For VMAT delivery to a moving target with a velocity of $V_{target}(t)$ in the BEV plane, the physical MLC leaf velocity $V_{L(R),k}^{phys,m}(t)$ in the *MLC frame of reference* is then given by:

$$V_{L(R),k}^{phys,m}(t) = V_k(g) \cdot \frac{dg}{dt} + V_{target}(t).$$

(9.B.4)

References

R. M. Abolfath and L. Papiez, General strategy for the protection of organs at risk in IMRT therapy of a moving body, *Med Phys* **36**, 3013–3017, 2009.

T. Bortfeld and S. Webb, Single-arc IMRT? *Phys Med Biol* **54**, N9–20, 2009.

L. Court, M. Wagar, R. Berbeco, A. Reisner, B. Winey, D. Schofield, D. Ionascu, A. M. Allen, R. Popple and T. Lingos, Evaluation of the interplay effect when using RapidArc to treat targets moving in the craniocaudal or right-left direction, *Med Phys* **37**, 4–11, 2010.

R. George, P. J. Keall, V. R. Kini, S. S. Vedam, J. V. Siebers, Q. Wu, M. H. Lauterbach, D. W. Arthur and R. Mohan, Quantifying the effect of intrafraction motion during breast IMRT planning and dose delivery, *Med Phys* **30**, 552–562, 2003.

E. Huang, L. Dong, A. Chandra, D. A. Kuban, I. I. Rosen, A. Evans and A. Pollack, Intrafraction prostate motion during IMRT for prostate cancer, *Int J Radiat Oncol Biol Phys Int J Radiat Oncol Biol Phys* **53**, 261–268, 2002.

P. J. Keall, G. S. Mageras, J. M. Balter, R. S. Emery, K. M. Forster, S. B. Jiang, J. M. Kapatoes, D. A. Low, M. J. Murphy, B. R. Murray, C. R. Ramsey, M. B. V. Herk, S. S. Vedam, J. W. Wong and E. Yorke, The management of respiratory motion in radiation oncology report of AAPM Task Group 76, *Med Phys* **33**, 3874–3900, 2006.

P. J. Keall, H. Cattell, D. Pokhrel, S. Dieterich, K. H. Wong, M. J. Murphy, S. S. Vedam, K. Wijesooriya and R. Mohan, Geometric accuracy of a real-time target tracking system with dynamic multileaf collimator tracking system, *Int J Radiat Oncol Biol Phys* **65**, 1579–1584, 2006.

R. McMahon, R. Berbeco, S. Nishioka, M. Ishikawa and L. Papiez, A real-time dynamic-MLC control algorithm for delivering IMRT to targets undergoing 2D rigid motion in the beam's eye view, *Med Phys* **35**, 3875–3888, 2008.

R. McMahon, L. Papiez and D. Rangaraj, Dynamic-MLC leaf control utilizing on-flight intensity calculations: A robust method for real-time IMRT delivery over moving rigid targets, *Med Phys* **34**, 3211–3223, 2007.

D. McQuaid and S. Webb, IMRT delivery to a moving target by dynamic MLC tracking: delivery for targets moving in two dimensions in the beam's eye view, *Phys Med Biol* **51**, 4819, 2006.

K. Otto, Volumetric modulated arc therapy: IMRT in a single gantry arc, *Med Phys* **35**, 310–317, 2008.

D. Palma, E. Vollans, K. James, S. Nakano, V. Moiseenko, R. Shaffer, M. McKenzie, J. Morris and K. Otto, Volumetric modulated arc therapy for delivery of prostate radiotherapy: comparison with intensity-modulated radiotherapy and three-dimensional conformal radiotherapy, *Int J Radiat Oncol Biol Phys* **72**, 996–1001, 2008.

L. Papiez and R. M. Abolfath, Variable beam dose rate and DMLC IMRT to moving body anatomy, *Med Phys* **35**, 4837–4848, 2008.

L. Papiez, R. McMahon and R. Timmerman, 4D DMLC leaf sequencing to minimize organ at risk dose in moving anatomy, *Med Phys* **34**, 4952–4956, 2007.

L. Papiez, The leaf sweep algorithm for an immobile and moving target as an optimal control problem in radiotherapy delivery, Mathematical and Computer Modelling **37**, 735–745, 2003.

L. Papiez, DMLC leaf-pair optimal control of IMRT delivery for a moving rigid target, *Med Phys* **31**, 2742–2754, 2004.

L. Papiez and D. Rangaraj, DMLC leaf-pair optimal control for mobile, deforming target, *Med Phys* **32**, 275–285, 2005.

L. Papiez, D. Rangaraj and P. Keall, Real-time DMLC IMRT delivery for mobile and deforming targets, *Med Phys* **32**, 3037–3048, 2005.

D. Rangaraj, G. Palaniswaamy and L. Papiez, DMLC IMRT delivery to targets moving in 2D in Beam's Eye View, *Med Phys* **35**, 3765–3778, 2008.

D. Rangaraj and L. Papiez, Synchronized delivery of DMLC intensity modulated radiation therapy for stationary and moving targets, *Med Phys* **32**, 1802–1817, 2005.

A. Sawant, R. Venkat, V. Srivastava, D. Carlson, S. Povzner, H. Cattell and P. Keall, Management of three-dimensional intrafraction motion through real-time DMLC tracking, *Med Phys* **35**, 2050–2061, 2008.

V. Srivastava, P. Keall, A. Sawant and Y. Suh, TU-C-M100J-06: Accurate Prediction of Intra-Fraction Motion Using a Modified Linear Adaptive Filter, *Med Phys* **34**, 2546, 2007.

B. Sun, D. Rangaraj, L. Papiez, S. Oddiraju, D. Yang and H. H. Li, Target tracking using DMLC for volumetric modulated arc therapy: A simulation study, *Med Phys* **37**, 6116–6124, 2010.

G. Tang, M. A. Earl, S. Luan, C. Wang, M. M. Mohiuddin and C. X. Yu, Comparing radiation treatments using intensity-modulated beams, multiple arcs, and single arcs, *Int J Radiat Oncol Biol Phys* **76**, 1554–1562, 2010.

S. Ulrich, F. Sterzing, S. Nill, K. Schubert, K. K. Herfarth, J. Debus and U. Oelfke, Comparison of arc-modulated cone beam therapy and helical tomotherapy for three different types of cancer, *Med Phys* **36**, 4702–4710, 2009.

W. F. Verbakel, J. P. Cuijpers, D. Hoffmans, M. Bieker, B. J. Slotman and S. Senan, Volumetric intensity-modulated arc therapy vs. conventional IMRT in head-and-neck cancer: a comparative planning and dosimetric study, *Int J Radiat Oncol Biol Phys* **74**, 252–259, 2009.

S. Webb, Motion effects in (intensity modulated) radiation therapy: a review, *Phys Med Biol* **51**, R403, 2006.

Q. J. Wu, F. F. Yin, R. McMahon, X. Zhu and S. K. Das, Similarities between static and rotational intensity-modulated plans, *Phys Med Biol* **55**, 33–43, 2010.

B. Y. Yi, S. Han-Oh, F. Lerma, B. L. Berman and C. Yu, Real-time tumor tracking with preprogrammed dynamic multileaf-collimator motion and adaptive dose-rate regulation, *Med Phys* **35**, 3955–3962, 2008.

10

Real-Time Motion Adaptation in Tomotherapy® Using a Binary MLC

10.1 Introduction 93
10.2 Binary MLC and TomoTherapy® Treatment System 93
10.3 Real-Time Motion Adaptation Strategies 94
    Motion-Adaptive Delivery • Motion-Adaptive Optimization
10.4 Simulations..97
    Synthetic Data • Clinical Data
10.5 System Integration and Experiments 99
10.6 Discussion..101
10.7 Conclusions...103
References..103

Weiguo Lu*

Mingli Chen

Carl J. Mauer

Gustavo H. Olivera

10.1 Introduction

The use of IMRT to sculpt the high-dose gradient region around the target and to spare the organ at risk is well established after decades of development (Webb 2006). However, the most important aspect of radiation therapy is locating the target and ensuring that the target is where it is supposed to be at the time of treatment. Although no one doubts that the dedicated IMRT can approach better dose distributions than conformal therapy for static targets, many investigators have demonstrated that small radiation field delivery with target motion can result in hot or cold spots in dose distributions (Yu et al. 1997). Motion remains one of the major challenges when applying IMRT to the thorax region in which respiratory motion dominates.

IMRT delivery can be classified as static beam IMRT, in which the radiation source is fixed during beam delivery, and rotational IMRT, in which the radiation source moves around the patient during delivery. Modulation of photon beam intensity is most commonly implemented through multileaf collimators (MLC) (Seco et al. 2001), which include (2D) field-shaping MLC and (1D) binary MLC (bMLC) (Yang et al. 1997; Balog et al. 1999; Sarkar et al. 2007). Both field-shaping MLC and bMLC can be used as static beam delivery or rotational beam delivery. Static beam IMRT includes step-and-shoot, dMLC-based, and topotherapy

(TomoDirect^SM) delivery mode; the former two use 2D MLC, while the third uses bMLC. Rotational IMRT includes intensity-modulated arc therapy (Yu et al. 2002; Yu 1997) and helical tomotherapy (TomoHelical^SM) (Mackie et al. 1995; Mackie et al. 1993); the former uses 2D MLC and the latter uses bMLC.

Many approaches regarding intrafractional motion management using 2D MLC, including breath-hold methods, gating methods, and dMLC tracking methods, are reported in other chapters of this book. Because of the unique nature of bMLC and the TomoTherapy® system, its motion management is expected to be different from that of conventional IMRT that is based on 2D MLC. In this chapter, we summarize strategies of real-time motion adaptation for bMLC-based IMRT, including TomoDirect^SM and TomoHelical^SM. We refer the readers to Lu (2008a, b) and Lu et al. (2009) for detailed algorithms and strategies.

10.2 Binary MLC and TomoTherapy® Treatment System

The source in the TomoTherapy® system (Figure 10.1) emits a 6-MV beam collimated using tungsten to keep the leakage dose outside of the treatment area very low, even for high monitor unit treatments. The source and collimation apparatus is mounted on a ring gantry, and the beam is shaped and modulated by 64 10-cm-thick binary MLC leaves (Figure 10.1). Each MLC leaf projects a width of about 0.6 cm at the isocenter plane and moves in or out of the field in 15–20 ms (transition time, from full open to full

---

* The techniques presented in this chapter are still "Work in Progress (WIP)" of TomoTherapy Inc.; no commercial product is available at the time of publication. Please send all correspondence to weiguolu@gmail.com.

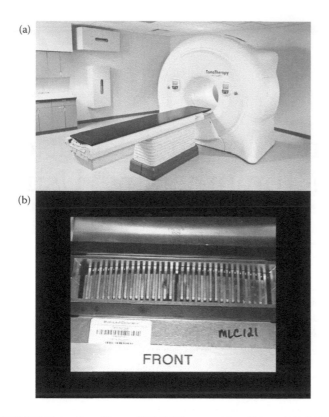

FIGURE 10.1 TomoTherapy® HiART® system (a) and binary MLC (b).

close or vice versa); it is individually controlled using a pneumatic system with no electrical motors with a leaf response latency of about 30 ms (TomoTherapy Inc. 2007). Radiation is either blocked or unblocked by a leaf—hence the name binary MLC.

The beam emanates from the collimation system in a "fan" shape of variable width and is used to both image and treat the patient in direct (TomoDirect^SM) or helical (TomoHelical^SM) fashion; the former uses a few fixed gantry angles and the latter uses continuous gantry rotation. Both TomoDirect^SM and TomoHelical^SM treatments follow a delivery plan which is optimized in the treatment planning procedure. The plan specifies the jaw width, the couch speed, and the sinogram (leaf open time for each projection). For TomoHelical^SM, the plan also includes information of the gantry speed. Individual leaf open time (or beamlet weight), which follows the planned sinogram, changes from projection to projection. Tens of thousands of overlapped, individually optimized beamlets are used in the treatment.

One key specification of real-time motion adaptation system is its system latency. System latency is the time between some status change (e.g., motion) and the action taken to compensate for the change. To accommodate system latency, prediction is often required. For motion-adaptive delivery strategies, system latency includes the motion detection time, leaf response time, and leaf transition time. Motion detection through external and/or internal surrogates can be updated as fast as 10–30 Hz, which means detection time is about 30 ms. The leaf response time of binary MLC is about 30 ms, and the leaf transition time is about 15–20 ms. Combining all these together, it results in a latency of 75 ms. On

the other hand, the current TomoTherapy delivery is projection-wised with the fastest projection rate of ~5 Hz (200 ms projection time), and our motion-adaptive delivery is also projection-wised, which means motion compensation is applied via the coming projections. Therefore, one half of the projection time (100 ms), the system's temporal resolution, should be added to the system latency; this renders the system latency about 175 ms, that is, we must predict the motion at least 175 ms ahead. Although long-term (e.g., >1 second) respiration prediction is very difficult, short-term prediction (e.g., 100–200 ms) can easily achieve sub-millimeter accuracy. The high projection rate of TomoTherapy delivery, in this sense, helps capture the instantaneous tumor position and makes the real-time motion compensation feasible.

10.3 Real-Time Motion Adaptation Strategies

Besides using 1D bMLC for intensity modulation, the basic TomoTherapy delivery has features that include fixed jaw width, fixed jaw position and jaw orientation, constant couch speed, and constant gantry rotation speed. The main advantage of such a system is its simplicity in both planning and delivery. However, such simplicity in a delivery system also poses some limitations in the cases of moving tumors (e.g., tumor motion due to respiration).

The conventional motion compensation that tracks the tumor either via 2D MLC or couch movement does not apply to the current TomoTherapy® system. In principle, tumor motion can be tracked by moving the couch or jaws so that the relative position of the tumor and treatment head stays the same as planned. But these would require changes in both hardware and control scheme of the current TomoTherapy system, which would be the scope of research in future. Novel approaches that are purely software-based were proposed by Lu (2008a, b) and Lu et al. (2009), which are briefly described here. These approaches include motion-adaptive delivery (MAD) that rearranges the planned sinogram to compensate for the tumor motion and motion-adaptive optimization (MAO) guided delivery that reoptimizes the leaf open time of the next projection. The following are the common features of MAD and MAO techniques:

1. Real-time motion compensation.
2. No breathing synchronization required.
3. Nearly 100% duty cycle.
4. Pure software solution and thus compliance with current hardware and work flow of the TomoTherapy® machine.
5. They apply to both TomoDirect^SM and TomoHelical^SM.

10.3.1 Motion-Adaptive Delivery

The MAD approach is a pure software solution. The idea of MAD is to deliver *the same radiation at the same position of the tumor, but not necessarily at the same time,* as planned.

This technique uses the planned leaf sequence but rearranges the projection indices and leaf indices. The implementation starts with a plan that is optimized via a regular TomoTherapy planning

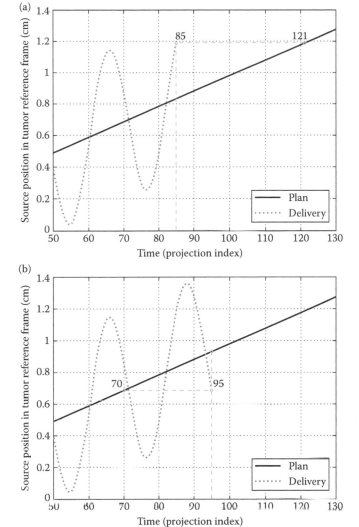

(a)

(b)

FIGURE 10.2 Illustration of longitudinal motion compensation of the MAD strategy. Both (a) and (b) show the source trajectories in the tumor reference frame (solid: planned; dashed: delivery, superposition of couch motion and tumor motion). (a) At the 85th projection, the source's position actually corresponds to the 121st projection. The MAD strategy will pick the 121st projection instead of using the planned 85th projection. (b) Similarly, at the 95th projection, the MAD strategy will pick the 70th projection, which the source's position actually corresponds to. (Reproduced from Lu, W., *Phys. Med. Biol.*, 53, 6491–511, 2008a. With permission.)

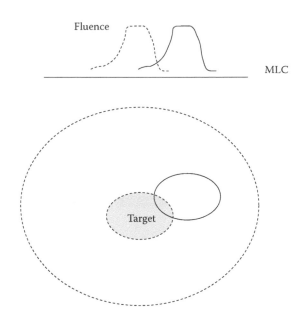

FIGURE 10.3 Illustration of transversal motion compensation of the MAD strategy. The dashed curves indicate the planned position and planned fluence. The solid curves indicate the delivery position and delivery fluence which is obtained by shifting the planned sinogram. (Reproduced from Lu, W., *Phys. Med. Biol.*, 53, 6491–511, 2008a. With permission.)

station. During planning, both the tumor and OAR are assumed static, and no motion margin is added. During delivery, the jaw width, couch speed, and gantry speed are all fixed as specified in the treatment plan. The leaf sequence is modified in real-time to compensate for the tumor motion. If the tumor is static, then the trajectory of the radiation source moving along the longitudinal direction with constant speed is a straight line in the tumor reference frame as illustrated by the solid line in Figure 10.2. If the tumor moves during treatment, then the trajectories of the radiation source is an oscillating curve, which is the superposition of

couch motion and tumor motion, as illustrated by the dashed line in Figure 10.2. Considering first longitudinal motion, which is usually more significant than transverse motion in the case of respiration, the MAD strategy picks the projection so that the right one according to the plan is delivered, as illustrated in Figure 10.2. Transverse motion comprises two components: the perpendicular and parallel components, relative to the beam direction. The perpendicular component can be compensated for by shifting the bMLC pattern (Figure 10.3) with adjustment for the cone effect. The parallel component can be taken care of using the inverse square correction. For rotational beam delivery (helical mode), the MAD strategy needs to pick projections that match the planned projection angle (Figure 10.4), besides matching the longitudinal position and correcting for the cone effect and inverse squared distance.

MAD is based on the assumption that every point of the moving tumor has approximately equal chance of being irradiated under a fixed jaw width, and this assumption is valid for periodic motion. Relatively regular respiratory motion is a good approximation of the periodic motion, and therefore a good candidate for the simple MAD technique. As for highly irregular motion or general form of motion, MAD could only be used to reduce the motion margin. It still leaves some dose heterogeneity in the tumor region. Some patient breath coaching will be helpful to maintain the regularity of respiration and therefore increase the applicability of the MAD technique. The suboptimal correction of MAD for very irregular motion can be improved by adopting real-time optimization and the negative feedback, self-correcting, closed-loop system, as implemented in MAO.

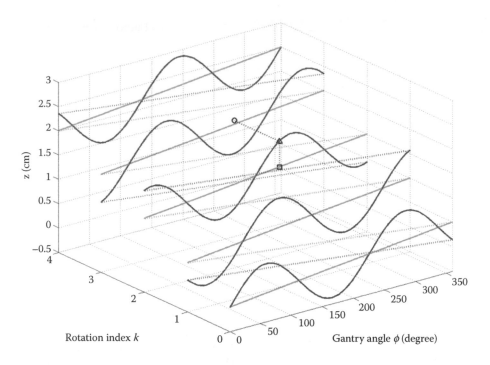

FIGURE 10.4 Illustration of longitudinal motion compensation of the MAD strategy for the helical mode. The solid straight lines and curved lines are the planned and delivery source trajectories, respectively, in the tumor reference frame. The delivery source trajectory is the superposition of couch motion and tumor motion. At the projection indicated by the square, the source is at the position indicated by the triangle. The MAD strategy delivers the projection indicated by the circle. (Reproduced from Lu, W., *Phys. Med. Biol.,* 53, 6513–31, 2008b. With permission.)

10.3.2 Motion-Adaptive Optimization

Radiotherapy treatment planning is based on information about the commissioning data of the delivery machine and the patient anatomy such as CT. Because the goal of treatment delivery is to reproduce the treatment plan as accurately as possible, the delivery procedure should be accurately modeled in treatment planning. However, real-time changes, such as tumor motion, are hard to be accurately modeled in advance. The state-of-the-art radiation delivery, including MAD, is an open-loop procedure. It tries to reproduce the planned procedures step by step but lacks the mechanism to deal with the error occurred and accumulated in each step. The real-time MAO guided radiotherapy changes the delivery scheme from a conventional open-loop system to a closed-loop system with negative feedback.

The MAD strategy works for relatively regular respiration, but renders suboptimal compensation for highly irregular motion. A more sophisticated strategy involving real-time optimization is developed in (Lu et al. 2009). This strategy—MAO—does not rely on the assumption of motion being relatively regular. Like the MAD strategy, MAO is based on an optimized plan—trying to reproduce the same. MAO can be divided into four major steps: (1) motion detection and prediction, (2) delivered dose accumulation, (3) future dose estimation, and (4) next projection optimization. Steps (2)–(4) are based on the knowledge of step (1). Its workflow is illustrated in Figure 10.5. MAO can be regarded as a closed-loop, negative feedback system that corrects

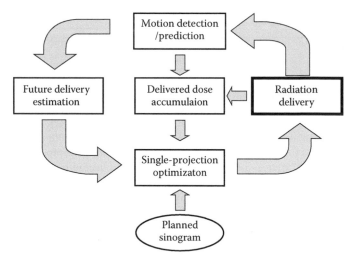

FIGURE 10.5 Workflow of MAO. MAO is a closed-loop, negative feedback system. The detected motion can be used for motion-encoded dose calculation to calculate the accumulated delivered dose and the predicted motion can be used to estimate future dose. The past (delivered dose) and future together with the planned dose are used to optimize the next projection. The optimized projection is sent to the machine to deliver radiation. The treatment beam can in turn be used to detect motion. (Reproduced from Lu, W., Chen, M., Ruchala, K.J., Chen, Q., Langen, K.M., Kupelian, P.A., and Olivera, G.H., *Phys. Med. Biol.,* 54, 4373–98, 2009. With permission.)

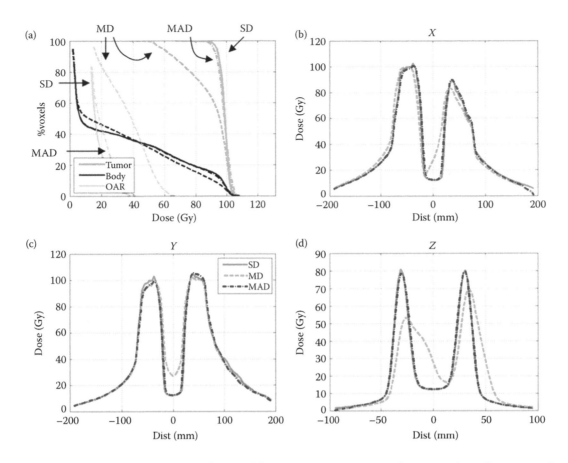

FIGURE 10.6 Comparison of DVHs and dose profiles for static delivery (SD, no tumor motion, the same as planned), motion without compensation delivery (MD) and motion-adaptive delivery (MAD). Figure (a) compares the DVHs of SD (solid), MD (dashed) and MAD (dash-dot). Figures (b), (c), and (d) show the dose profiles along the lateral (*X*), AP (*Y*) and SI (*Z*) directions, respectively. The solid line indicates the SD profile, the dashed line indicates the MD profiles, and the dash-dot line indicates the MAD profiles. All profiles go through the center in the tumor reference frame. (Reproduced from Lu, W., *Phys. Med. Biol.*, 53, 6513–31, 2008b. With permission.)

cumulative errors. We implemented and tested the MAO technique using a TomoTherapy® research system.

Both MAD and MAO techniques are easy to implement in real time because of the following two aspects:

1. Unlike dynamic MLC-based tracking, MAO/MAO does not need the whole tumor motion trajectory or tumor velocity. It only requires the instantaneous tumor position, which greatly simplifies the system implementation.
2. MAD reuses the planned leaf sequence by rearranging (out of order execution) the projection index and leaf index. MAO optimizes only the next projection. Their calculations can be accomplished in real time.

10.4 Simulations

10.4.1 Synthetic Data

The first simulation used a plan that was created on the MVCT image of a mini cheese phantom of 20 cm in diameter with a C-shaped target enclosing a ball OAR and both positioned at the isocenter. The delivery used the jaw width of 1.05 cm, pitch of

0.3, and gantry period of 10 seconds, which corresponds to the fastest gantry speed of the current TomoTherapy® machine and the worst-case scenario of the TomoTherapy delivery error for respiratory motion without correction. Motion of Lujan-type regular respiration with the period of 4.3 seconds and amplitude of 2 cm was simulated in both *X* (lateral) and *Z* (longitudinal) directions. We calculated planned dose (SD dose), delivered dose with tumor motion (MD dose), and delivered dose with motion correction (MAD dose) and compared their respective dose volume histograms (DVHs). Figure 10.6 compares DVHs (the top left panel) and dose profiles along the *X*, *Y*, and *Z* directions (other panels). We can see that the planned SD dose conforms well to the C-shaped target and avoids the ball OAR. The delivered dose deviated significantly from the planning objective due to large motion amplitude. However, the MAD strategy compensated well for this kind of motion. The resulting MAD dose distribution matched well with the static dose, as illustrated by DVHs and dose profiles (Figure 10.6).

In the second simulation, we used 1D data to study the effects of longitudinal motion. Because the motion was only longitudinal, the results applied to both TomoDirect^{SM} and TomoHelical^{SM}

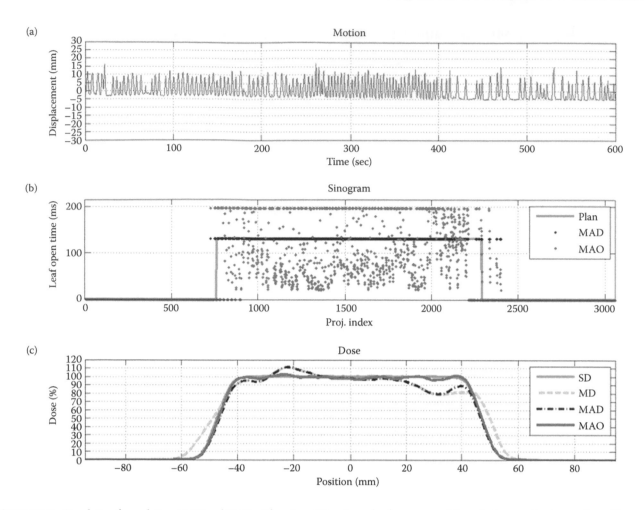

FIGURE 10.7 Simulation for real-time MAD and MAO with a given 1D rectangular fluence map. (a) respiratory motion trace of a real lung cancer patient. (b) planned sinogram (solid line), MAD sinogram (black dot) and MAO sinogram (gray dot) for the motion in (a). (c) comparison of static dose (gray solid line, delivered by the planned sinogram for the stationary tumor), MD dose (dashed line, delivered by the planned sinogram for the moving tumor), MAD dose (black dot-dashed line, delivered by the MAD sinogram for the moving tumor) and MAO dose (darker gray solid line, delivered by the MAO sinogram for the moving tumor). (Reproduced from Lu, W., Chen, M., Ruchala, K.J., Chen, Q., Langen, K.M., Kupelian, P.A., and Olivera, G.H., *Phys. Med. Biol.,* 54, 4373–98, 2009. With permission.)

delivery modes. We compared results of the MAD and MAO guided deliveries. A uniform dose distribution was intended for an 8-cm-long target. The projection time was 196 ms (=10 seconds per rotation/51 projections per rotation). The planned sinogram was a simple rectangle-shaped profile, corresponding to the leaf open time of 131 ms (=196 ms/MF) in the tumor region and 0 elsewhere. Here, MF (=1.5) denotes the modulation factor used in the simulation.

The result is shown in Figure 10.7. The top panel is the longitudinal motion trace, the middle panel is the planned and delivery sinograms, and the bottom panel compares the doses of SD, MD, MAD, and MAO. The top panel shows a very irregular respiration trace from a lung cancer patient with the peak-to-peak amplitude scaled to 2 cm. Just like any IMRT technique, the combination of the large amplitude of very irregular respiration, small field size (1.05 cm) and fast gantry rotation (10 seconds) of TomoTherapy delivery resulted in high tumor dose nonuniformity in the MD dose as illustrated in the bottom panel of

Figure 10.7. The MAD dose again matched the SD dose well on the tumor boundary, but it had the same nonuniformity as the MD dose in the inner region of the tumor. On the other hand, the MAO method reduced edge blurring and tumor dose nonuniformity effects due to irregular respiration to within 3% difference from the SD dose.

10.4.2 Clinical Data

As an example, we studied retrospectively a lung cancer case with a tumor size of approximately 2 cm in the inferior part of the lung. A TomoTherapy plan was optimized using TomoTherapy® HiArt® II TPS with the jaw width of 2.5 cm and pitch of 0.3. The optimization used the GTV as the target without any motion margin. The measured spirometer signals were 1D only and provided the relative amplitude and phase information. We scaled the signal amplitude so that the range of its lower 10% to upper 10% was 3 cm in the SI direction, 2 cm in the AP direction and

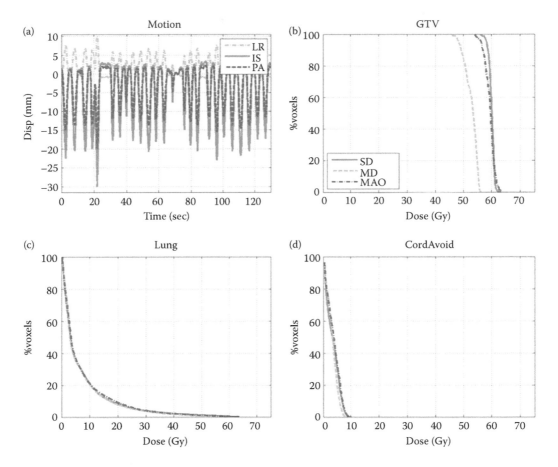

FIGURE 10.8 DVH comparisons of different delivery methods—planned delivery without motion (SD, solid line), planned delivery with motion (MD, dashed line), and MAO guided delivery (MAO, dot-dashed line) for a lung cancer patient under TomoTherapy treatment. (a) respiration motion trace with the peak-to-peak amplitude of about 30 mm in the SI direction, 20 mm in the AP direction, and 10 mm in the LR direction. Figures (b), (c), and (d) compare DVHs for GTV, the lung, and the spinal cord with an avoidance region. (Reproduced from Lu, W., Chen, M., Ruchala, K.J., Chen, Q., Langen, K.M., Kupelian, P.A., and Olivera, G.H., *Phys. Med. Biol.*, 54, 4373–98, 2009. With permission.)

1 cm in the LR direction. This motion range is close to the maximum respiratory motion reported in the literature. The motion traces are shown in the top left panel of Figure 10.8. The treatment planning was based on the maximum expiration phase, which was used as the reference phase for all dose mappings. We calculated the MD and MAO dose and compared them with the SD dose using DVHs (Figure 10.8). There were significant cold spots in the MD dose. Note that this was an extreme case, as expected for any IMRT, because the GTV was only about 2 cm and the motion was as large as 3 cm. In addition, without motion margin in the treatment plan, it is expected that significant cold spots will show up because the tumor may move out of the radiation field. But even with no motion margin, the DVHs of MAO matched the planned DVHs (SD) very well with negligible cold spots. The results indicated that the real-time MAO technique is an effective way to reduce the margin for treatment of a small lung tumor that undergoes significant respiratory motion.

Figure 10.9 shows the results of a retrospective study that applies MAO to a prostate case. The prostate moved between 3 mm and 6 mm in the SI direction for a significant amount of time, categorized as a medium-range prostate motion. Both hot and cold spots appeared in the prostate DVH of MD compared with that of SD, because the tumor moved both upstream (against the radiation source motion) and downstream (following the radiation source motion) in the SI direction, as illustrated by the SI motion trace in the top left panel of Figure 10.9. The upstream motion caused parts of the prostate receiving lower-than-planned dose, while the downstream motion caused other parts higher-than-planned dose. The MAO technique was able to compensate for both motions as illustrated in the DVH plots. The prostate DVH of MAO approached that of SD well. In addition, the hot spots in the bladder were corrected by MAO.

10.5 System Integration and Experiments

We implemented both MAD and MAO algorithms in C++ programming language and integrated it with a TomoTherapy delivery system in the research bunker of TomoTherapy Inc. Information about "system integration and experimental validation" was presented in ASTRO2008 (Lu et al. 2008b). The whole algorithm of real time MAD/MAO including motion

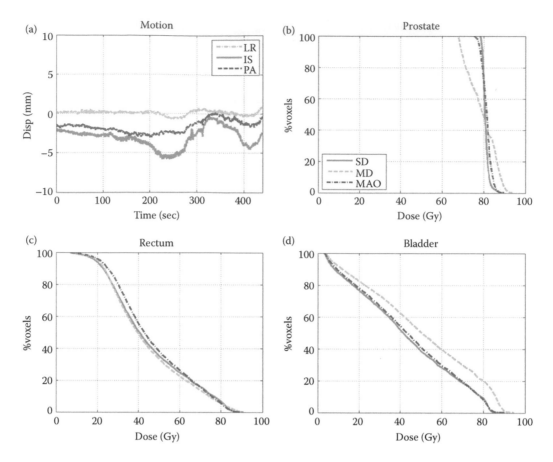

FIGURE 10.9 DVH comparisons of different delivery methods—SD (solid), MD (dashed), and MAO (dot-dashed) for a prostate cancer patient undergoing TomoTherapy treatment. Figure (a) shows prostate motion traces in the LR, IS, and PA directions, respectively. Figures (b), (c), and (d) compare DVHs for the prostate, rectum, and bladder, respectively. (Reproduced from Lu, W., Chen, M., Ruchala, K.J., Chen, Q., Langen, K.M., Kupelian, P.A., and Olivera, G.H., *Phys. Med. Biol.*, 54, 4373–98, 2009. With permission.)

prediction, delivered dose accumulation, future dose estimation, and single projection optimization took less than 100 ms per projection, which was sufficient for real-time delivery modification in TomoTherapy. The integrated system includes the following:

1. A workstation with MAD/MAO software
2. A tracking camera
3. A 4D phantom consisting of a three-axis motorized stage and an attached mini-cheese phantom (Figure 10.10)

The whole integrated system is illustrated in Figure 10.11.

The 4D phantom was placed on the couch with the axes aligned with the lasers. Films and ion chambers were placed in the mini-cheese phantom to measure dose. An infrared tracking marker was rigidly attached to the 4D phantom frame. The MAO workstation ran an application to drive the motion of the 4D Phantom.

The tracking marker's position was determined by the camera and was transmitted to the MAD/MAO workstation. The MAD/MAO workstation (shown in Figure 10.11) received the initial sinogram and patient information from the operator station of the TomoTherapy® research system. The onboard computer

notified the MAD/MAO workstation the current projection. These communications were handled over Ethernet, with the exception of the camera which used a serial USB connection. Note that although we used camera as surrogate in our experiments, the algorithm and system itself is independent of any surrogate.

During the delivery procedure, the MAD/MAO workstation performed motion correction by running MAD/MAO which calculated a new sinogram entry (for the next projection) and the result was sent to the onboard computer.

Figure 10.12 illustrates film measurements of SD, MD, and MAO delivery mode for using this integrated system. The motion trace (shown in top half) that drove the phantom was from the spirometer signal of a lung cancer patient. The treatment plan was to deliver uniform dose to a saddle-shaped target. The static mode, as shown in the top-right film, shows a perfect saddle-shaped dose distribution. The MD (motion without correction, lower-left film) mode shows significant motion artifacts, including blurs and ripples, in dose distribution. The MAO (motion with correction, lower-right film) compensated for the motion and dose distribution appears similar to the static mode.

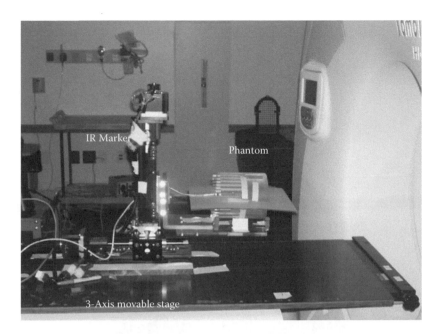

FIGURE 10.10 4D phantom. A three-axis moveable stage is used to drive a cheese phantom that simulates target motion. An RF-camera system is used to track motion surrogated by infra-red marker. Coronal films are used to measure dose of the static mode (SD), motion without correction mode (MD) and MAO mode (MAO).

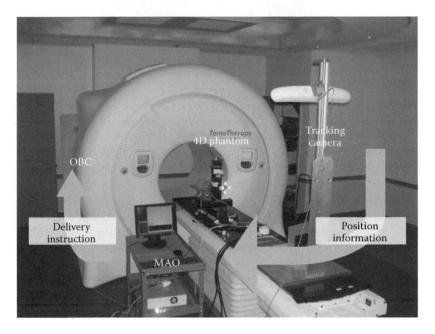

FIGURE 10.11 Integration of the MAO apparatus with a TomoTherapy® research system. In this integration, a tracking camera is used to detect target motion in real time. The target position information is sent to a workstation that runs the MAO algorithm, which outputs the delivery instruction to the onboard computer that controls the bMLC to modulate the radiation beam.

10.6 Discussion

IMRT delivery via binary MLC is simpler than that via conventional field-shaping MLC, and its implementation is easy and reliable. The basic TomoDirect and TomoHelical delivery system has features such as fixed jaw widths, fixed jaw positions and orientations, constant couch speed, and 1D intensity modulation.

However, such simplicity causes limitations in the cases of moving tumors (e.g., tumor motion resulting from respiratory motion). Hence, a motion compensation technique needs to be developed.

Just like the beam-gating- and beam-tracking-based methods, the presented motion adaptation techniques rely on the information of precise tumor position in real-time. Respiration

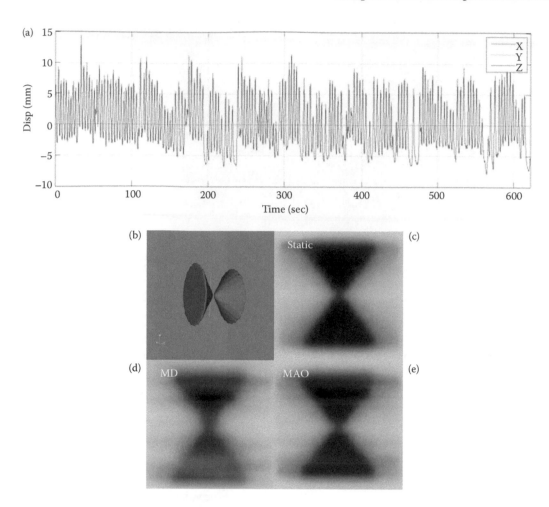

FIGURE 10.12 Illustration of the motion trace (a), target (b) and coronal film measurement of dose distribution for static (c), motion without correction (d) and motion with correction (e) delivery mode.

monitoring techniques that are available include surrogate-based methods (Simon et al. 2005; Kimura et al. 2004; Garcia et al. 2002; Zhang et al. 2003), internal fiducial-based methods (Shirato et al. 2000; Chen et al. 2001; Seppenwoolde et al. 2002; Rietzel et al. 2004; Keall et al. 2004), hybrids (Schweikard et al. 2004) and treatment beam methods (Lu et al. 2006b; Meyer et al. 2006). The surrogate methods rely on the correlation between external surrogates, such as skin markers (Ozhasoglu and Murphy 2002) or lung air flow (Simon et al. 2005; Lu et al. 2005; Kimura et al. 2004; Garcia et al. 2002; Zhang et al. 2003), and internal motion of the structures of interest (Koch et al. 2004; Low et al. 2005; Wu et al. 2004). Such correlations are not always reliable (Koch et al. 2004; Hoisak et al. 2004), especially for irregular breathing. The internal fiducial-based methods directly detect the position of internal fiducial markers by means of online fluoroscopic imaging (Shirato et al. 2000; Chen et al. 2001; Seppenwoolde et al. 2002; Rietzel et al. 2004), electromagnetic signal detection (Willoughby et al. 2006; Kupelian et al. 2007), or radioactive signal detection (Shchory et al. 2008). Although the motion-adaptation techniques presented in this chapter are independent of motion-detection methods, we think that internal fiducial-based methods or the treatment beam-based methods would be more

suitable than surrogate-based methods for detecting tumor position in real time because of the direct tumor position measurement of the former two.

The MAO guided delivery is a strategy to deal with irregular respiration, as well as general form of motion. For regular respiration, the MAD strategy performs very well. In our simulation, MAD dose matched the static dose far less than 3 mm and 3% criteria. However, for irregular motion that fails the assumption of equal chance of irradiation, a more sophisticated strategy in MAO needs to be applied.

The framework of MAO can be extended to a more general scheme, real-time adaptive radiation therapy (ART) (Yan et al. 1997a, b). ART generally refers to the concepts of using feedback during the course of radiation therapy to improve future treatment. Feedback can be used for off-line adaptive processes (Lu et al. 2006a; Yan et al. 2005) or online (Olivera et al. 2005; Ahunbay et al. 2008; Wu et al. 2008) processes. Off-line ART refers to processes when the patient is not being treated, such as in between treatment fractions, whereas online ART refers to processes when the patient is on the treatment couch right before delivery of the treatment beam. Both off-line and online ART are to compensate for interfractional changes. Real-time

ART, however, is to correct intrafractional, or real-time, generated errors caused due to either patient motion or random machine variations, such as linac output changes, leaf open errors, gantry rotation errors, and couch motion errors. Real-time ART imposes a high demand on error detection and system response. Real-time dose reconstruction will also be a critical component for real-time ART. The real-time optimization workflow described in this paper should be able to accomplish real-time ART for TomoTherapy delivery, provided reliable implementation of motion detection, delivery verification, and dose reconstruction (Kapatoes et al. 2001) can be established.

The MAD/MAO technique still falls in the category of "compensation" that regards motion as an "error" to be corrected. Motion is a challenge, yet motion is also a chance. Just as we can take advantage of interfractional variations of the tumor–OAR configuration to achieve better therapeutic gain through "Adaptive Fractionation Therapy" (Lu et al. 2008a; Chen et al. 2008), we can potentially take advantage of intrafraction motion to achieve a "better-than-planned" delivery. This possibility was demonstrated by Papiez's group (Papiez et al. 2007) using 4D DMLC delivery to minimize the OAR dose. We believe that a more advanced real-time optimization scheme should be able to offer a superior delivery, which is not achievable via any plan based on static delivery. Such scheme may require more sophisticated algorithms and a powerful computer. It definitely needs further exploration.

In the current MAD/MAO strategies, we assumed that tumor motion was rigid body shift and OAR moved the same way as the tumor. This is also a basic assumption for DMLC tracking techniques used in conventional IMRT. This is a strong assumption for a small tumor and for OAR close to the tumor. Motion compensation has more significant effect on a small tumor than on a large tumor, and the case when OAR is close to the tumor is usually of more concern than that when OAR is far away from the tumor. Therefore, we could regard such assumptions valid for the majority of cases when motion is a concern.

For other tumor motion, such as rotation or general deformation, it requires different formulas to calculate motion-encoded beamlets. The problem is then how to accurately delineate the deformation in real time. This may involve an ultrafast deformable registration algorithm (Samant et al. 2008; Lu et al. 2004) or the use of precalculated deformation maps that are based on some 4D images such as 4DCT (Lu et al. 2006b).

10.7 Conclusions

Intrafraction target motion confronts IMRT delivery where precise target and RAR positions are assumed. Simulations and experiments, including the worst-case scenarios, validated our motion adaptation techniques. The adaptation without optimization (MAD) is applicable for relatively regular or mild irregular respirations. The adaptation with optimization (MAO) is suitable for regular or irregular breathing-type motion as well as drift and random motion of prostate patients.

In the tests, the delivered dose conformed well to the target and a significant margin reduction was achieved, provided that accurate, real-time tumor localization was available. The presented strategies for delivering static (TomoDirect^SM) and rotational (TomoHelical^SM) beams to compensate for moving tumors do not require extra hardware or control of the current TomoTherapy® system.

As in any forms of IMRT, without employing motion adaptation strategies, significant portions of the tumor volume could be underdosed because we did not include motion margin in planning. With motion adaptation strategies, the delivered dose could well achieve the planning objective. In other words, motion adaptation strategies can be used to reduce motion margins significantly in TomoTherapy delivery.

The system integration for real-time motion adaptive TomoTherapy delivery is still a work-in-progress project. No commercial product is available up to date. The presented motion adaption strategies and results are quite preliminary. They are based on current TomoTherapy delivery system that are categorized as fixed jaw width and fixed couch speed. Dynamic jaw and dynamic couch (DJDC) TomoTherapy delivery (Sterzing et al. 2009; Chen et al. 2009), which is under development by TomoTherapy Inc. DJDC opens a new paradigm for IMRT treatment, and it will certainly change the strategy for motion management. With the capability of the dynamic jaw and dynamic couch, a more effective real-time motion management technique that combines jaw and couch modulation with real-time MAO may become feasible.

References

Ahunbay E E, Peng C, Chen G P, Narayanan S, Yu C, Lawton C, and Li X A (2008) An on-line replanning scheme for interfractional variations *Med Phys*, **35**, 3607–15.

Balog J P, Mackie T R, Wenman D L, Glass M, Fang G and Pearson D (1999) Multileaf collimator interleaf transmission *Med Phys*, **26**, 176–86.

Chen M, Lu W, Chen Q, Ruchala K and Olivera G (2008) Adaptive Fractionation Therapy—II Biological Effective Dose *Phys Med Biol*, **53**, 5513–25.

Chen Q S, Weinhous M S, Deibel F C, Ciezki J P and Macklis R M (2001) Fluoroscopic study of tumor motion due to breathing: Facilitating precise radiation therapy for lung cancer patients *Med Phys*, **28**, 1850–6.

Chen Y, Lu W, Chen M, Chen Q, Schnarr E, Reitz G, Henderson D, Ruchala K J, and Olivera G (2009) Dynamic Delivery Techniques for Helical Tomotherapy Treatment *Int J Radiat Oncol Biol Phys*, **75**, S122.

Garcia R, Oozeer R, Le Thanh H, Chastel D, Doyen J C, Chauvet B, and Reboul F (2002) [Radiotherapy of lung cancer: The inspiration breath hold with spirometric monitoring] *Cancer Radiother*, **6**, 30–8.

Hoisak J D, Sixel K E, Tirona R, Cheung P C and Pignol J P (2004) Correlation of lung tumor motion with external surrogate indicators of respiration *Int J Radiat Oncol Biol Phys*, **60**, 1298–306.

Kapatoes J M, Olivera G H, Ruchala K J, Smilowitz J B, Reckwerdt P J, and Mackie T R (2001) A feasible method for clinical delivery verification and dose reconstruction in tomotherapy *Med Phys*, **28**, 528–42.

Keall P J, Todor A D, Vedam S S, Bartee C L, Siebers J V, Kini V R, and Mohan R (2004) On the use of EPID-based implanted marker tracking for 4D radiotherapy *Med Phys*, **31**, 3492–9.

Kimura T, Hirokawa Y, Murakami Y, Tsujimura M, Nakashima T, Ohno Y, Kenjo M, Kaneyasu Y, Wadasaki K, and Ito K (2004) Reproducibility of organ position using voluntary breath-hold method with spirometer for extracranial stereotactic radiotherapy *Int J Radiat Oncol Biol Phys*, **60**, 1307–13.

Koch N, Liu H H, Starkschall G, Jacobson M, Forster K, Liao Z, Komaki R, and Stevens C W (2004) Evaluation of internal lung motion for respiratory-gated radiotherapy using MRI: Part I—Correlating internal lung motion with skin fiducial motion *Int J Radiat Oncol Biol Phys*, **60**, 1459–72.

Kupelian P, Willoughby T, Mahadevan A, Djemil T, Weinstein G, Jani S, Enke C, Solberg T, Flores N, Liu D, Beyer D, and Levine L (2007) Multi-institutional clinical experience with the Calypso System in localization and continuous, real-time monitoring of the prostate gland during external radiotherapy *Int J Radiat Oncol Biol Phys*, **67**, 1088–98.

Low D A, Parikh P J, Lu W, Dempsey J F, Wahab S H, Hubenschmidt J P, Nystrom M M, Handoko M, and Bradley J D (2005) Novel breathing motion model for radiotherapy *Int J Radiat Oncol Biol Phys*, **63**, 921–9.

Lu W (2008a) Real-time Motion-Adapted-Delivery(MAD) using Binary MLC -I. static beam (topotherapy) delivery *Phys Med Biol*, **53**, 6491–511.

Lu W (2008b) Real-time Motion-Adapted-Delivery(MAD) using Binary MLC -II. rotational beam (tomotherapy) delivery *Phys Med Biol*, **53**, 6513–31.

Lu W, Chen M, Chen Q, Ruchala K, and Olivera G (2008a) Adaptive Fractionation Therapy—I. Basic Concept and Strategy *Phys Med Biol*, **53**, 5495–5511.

Lu W, Chen M, Olivera G H, Ruchala K, and Mackie T R (2004) Fast free-form deformable registration via calculus of variations *Phys Med Biol*, **49**, 3067–87.

Lu W, Chen M, Ruchala K J, Chen Q, Langen K M, Kupelian P A, and Olivera G H (2009) Real-time motion-adaptive-optimization (MAO) in TomoTherapy. *Phys Med Biol*, **54**, 4373–98.

Lu W, Low D A, Parikh P J, Nystrom M M, El Naqa I M, Wahab S H, Handoko M, Fooshee D, and Bradley J D (2005) Comparison of spirometry and abdominal height as four-dimensional computed tomography metrics in lung *Med Phys*, **32**, 2351–7.

Lu W, Mauer C, Chen M, Ruchala K, Zhang J, Lucas D, Chen Q, and Olivera G (2008b) Experimental Validation for Real-time Motion Adapted Optimization (MAO) Guided Delivery *Int J Radiat Oncol Biol Phys (ASTRO2008)*, **72**, s26.

Lu W, Olivera G H, Chen Q, Ruchala K J, Haimerl J, Meeks S L, Langen K M, and A. K P (2006a) Deformable Registration of the Planning Image (KVCT) and the Daily Images (MVCT) for Adaptive Radiation Therapy *Phys Med Biol*, **51**, 4351–74.

Lu W, Ruchala K J, Chen M-L, Chen Q, and Olivera G H (2006b) Real-Time Respiration Monitoring using the Radiotherapy Treatment Beam and Four-Dimensional Computed Tomography (4DCT)-A Conceptual Study *Phys. Med. Biol.*, **51**, 4469–95.

Mackie T R, Holmes T W, Reckwerdt P J, and Yang J (1995) Tomotherapy: Optimized planning and delivery of radiation therapy *Int. J. Imaging Syst. Technol.*, **6**, 43–55.

Mackie T R, Holmes T W, Swerdloff S, Reckwerdt P J, Deasy J O, Yang J, Paliwal B, and Kinsella T (1993) Tomotherapy: A new concept for the delivery of dynamic conformal radiotherapy *Medical Physics*, **20**, 1709–19.

Meyer J, Richter A, Baier K, Wilbert J, Guckenberger M, and Flentje M (2006) Tracking moving objects with megavoltage portal imaging: A feasibility study *Medical Physics*, **33**, 1275–80.

Olivera G H, Mackie T R, Ruchala K, Lu W, Jeswani S, Aoyama H, Kapatoes J, and Chen Q (2005) Helical tomotherapy process *Japanese Journal of Medical Physics*, **25**, 39–63.

Ozhasoglu C and Murphy M J (2002) Issues in respiratory motion compensation during external-beam radiotherapy *Int J Radiat Oncol Biol Phys*, **52**, 1389–99.

Papiez L, McMahon R, and Timmerman R (2007) 4D DMLC leaf sequencing to minimize organ at risk dose in moving anatomy *Med Phys*, **34**, 4952–6.

Rietzel E, Rosenthal S J, Gierga D P, Willet C G, and Chen G T (2004) Moving targets: Detection and tracking of internal organ motion for treatment planning and patient set-up *Radiother Oncol*, **73 Suppl 2**, S68–S72.

Samant S S, Xia J, Muyan-Ozcelik P, and Owens J D (2008) High performance computing for deformable image registration: Towards a new paradigm in adaptive radiotherapy *Med Phys*, **35**, 3546–53.

Sarkar V, Lin L, Shi C, and Papanikolaou N (2007) Quality assurance of the multileaf collimator with helical tomotherapy: Design and implementation *Med Phys*, **34**, 2949–56.

Schweikard A, Shiomi H and Adler J (2004) Respiration tracking in radiosurgery *Med Phys*, **31**, 2738–41.

Seco J, Evans P M, and Webb S (2001) Analysis of the effects of the delivery technique on an IMRT plan: Comparison for multiple static field, dynamic and NOMOS MIMiC collimation *Phys Med Biol*, **46**, 3073–87.

Seppenwoolde Y, Shirato H, Kitamura K, Shimizu S, van Herk M, Lebesque J V, and Miyasaka K (2002) Precise and real-time measurement of 3D tumor motion in lung due to breathing and heartbeat, measured during radiotherapy *Int J Radiat Oncol Biol Phys*, **53**, 822–34.

Shchory B T, Schifter D, Lichtman R, Neustadter D, and Corn B (2008) Static and Dynamic Tracking Accuracy of a Novel Radioactive Tracking Technology for Target Localization and Real Time Tracking in Radiation Therapy *Med Phys*, **36**, 2719.

Shirato H, Shimizu S, Kunieda T, Kitamura K, van Herk M, Kagei K, Nishioka T, Hashimoto S, Fujita K, Aoyama H, Tsuchiya K, Kudo K, and Miyasaka K (2000) Physical aspects of a real-time tumor-tracking system for gated radiotherapy *Int J Radiat Oncol Biol Phys*, **48**, 1187–95.

Simon L, Giraud P, Servois V, and Rosenwald J C (2005) Lung volume assessment for a cross-comparison of two breathing-adapted techniques in radiotherapy *Int J Radiat Oncol Biol Phys,* **63,** 602–9.

Sterzing F, Uhl M, Schubert K, Sroka-Perez G, Chen Y, Lu W, Mackie R, Debus J, Herfarth K, and Oliveira G (2010) Dynamic jaws and dynamic couch in helical tomotherapy *Int J Radiat Oncol Biol Phys,* **76(4),** 1266–73.

TomoTherapy Inc (2007) *The Beamlet,* http://www.tomotherapy.com/beamlet/beamlet011/.

Webb S (2006) Motion effects in (intensity modulated) radiation therapy: A review *Phys Med Biol,* **51,** R403–R425.

Willoughby T R, Kupelian P A, Pouliot J, Shinohara K, Aubin M, Roach M, III, Skrumeda L L, Balter J M, Litzenberg D W, Hadley S W, Wei J T, and Sandler H M (2006) Target localization and real-time tracking using the Calypso 4D localization system in patients with localized prostate cancer *Int J Radiat Oncol Biol Phys,* **65,** 528–34.

Wu H, Sharp G C, Salzberg B, Kaeli D, Shirato H, and Jiang S B (2004) A finite state model for respiratory motion analysis in image guided radiation therapy *Phys Med Biol,* **49,** 5357–72.

Wu Q J, Thongphiew D, Wang Z, Mathayomchan B, Chankong V, Yoo S, Lee W R, and Yin F F (2008) On-line re-optimization of prostate IMRT plans for adaptive radiation therapy *Phys Med Biol,* **53,** 673–91.

Yan D, Lockman D, Martinez A, Wong J, Brabbins D, Vicini F, Liang J, and Kestin L (2005) Computed tomography guided management of interfractional patient variation *Semin Radiat Oncol,* **15,** 168–79.

Yan D, Vicini F, Wong J, and Martinez A (1997a) Adaptive radiation therapy *Phys Med Biol,* **42,** 123–32.

Yan D, Wong J, Vicini F, Michalski J, Pan C, Frazier A, Horwitz E, and Martinez A (1997b) Adaptive modification of treatment planning to minimize the deleterious effects of treatment setup errors *Int J Radiat Oncol Biol Phys,* **38,** 197–206.

Yang J N, Mackie T R, Reckwerdt P, Deasy J O, and Thomadsen B R (1997) An investigation of tomotherapy beams delivery *Medical Physics,* **24,** 425–36.

Yu C (Ed.) (1997) *Intensity Modulated Arc therapy: A new method for delivering conformal radiation therapy,* Advance Medical Publishing, Durango, Colorado.

Yu C X, Jaffray D A, and Wong J W (1997) The effects of intra fraction organ motion on the delivery of dynamic intensity modulation *Phys. Med. Biol.,* **43,** 91–104.

Yu C X, Li X A, Ma L, Chen D, Naqvi S, Shepard D, Sarfaraz M, Holmes T W, Suntharalingam M, and Mansfield C M (2002) Clinical implementation of intensity-modulated arc therapy *Int J Radiat Oncol Biol Phys,* **53,** 453–63.

Zhang T, Keller H, O'Brien M J, Mackie T R, and Paliwal B (2003) Application of the spirometer in respiratory gated radiotherapy *Med Phys,* **30,** 3165–71.

Combination of LINAC with 1.5 T MRI for Real-Time Image Guided Radiotherapy

Jan J.W. Lagendijk

Bas W. Raaymakers

Marco van Vulpen

11.1 Introduction 107
11.2 Design Magnetic Resonance Linac 107
11.3 Status...108
11.4 Dosimetry and Treatment Planning.................................108
11.5 Clinical Impact...111
References...112

11.1 Introduction

Magnetic Resonance Imaging (MRI) has huge potential for radiotherapy treatment simulation, tumor characterization, and treatment response assessment. The soft tissue visualization of MRI is superior to that of other imaging modalities for almost all tumor and normal tissues. As an example, for prostate tumors, MRI is already the gold standard for delineation of the prostate (Fütterer et al. 2008), while dynamic contrast enhanced imaging (DCE-MRI) and diffusion weighted imaging (DWI) are used to determine the exact tumor position inside the prostate (Groenendaal et al. 2010; Jackson et al. 2009; Korporaal et al. 2010). (High field) MR spectroscopy is investigated for tumor characterization (Kurhanewicz and Vigneron 2008; Klomp et al. 2009). Lymph node involvement can be judged and tracked by using USPIOs (Barentsz et al. 2007) in combination with DWIBS (Klerkx et al. 2009). Recently, it was found that both DCE-MRI and DWI are very well suited to determine whether and where tumors recur inside the prostate making focal salvage therapy possible (Moman et al. 2010).

MRI offers even more. MRI allows multiple and high temporal resolution cine imaging and thus may be used to supply the information for adaptive radiotherapy. Tumor regression can easily be visualized (van de Bunt et al. 2008) while cine MRI gives information of breathingrelated movements and tumor stability (Ghilezan et al. 2005; Kirilova et al. 2008; Kerkhof et al. 2009). This huge potential of MRI can be safely exploited in real time without any additional radiation dose to the patient. This chapter describes how we can employ this potential and describes the use of an 1.5 T MRI integrated with a 6 MV linear accelerator for online tumor targeting and treatment response assessment. This MRI-Linac (MRL) must provide the ultimate soft tissue targeting system by facilitating real-time visualization of the tumor and its normal tissue surroundings during the actual treatment itself; this includes all irregular breathing related movements. We will show that the diagnostic quality 1.5T MRI allows the use of MRI cine and navigators for tracking and also the use of DWI for characterization, all during radiation beam on.

We describe how the expected difficulties such as magnetic distortion of the linac by the MRI, B0 magnetic distortion of the MRI by the linac, all potential RF distortion, distorted dose distribution in magnetic fields, absolute dosimetry in magnetic fields, geometrical accuracy, calibration of the world coordinate system, and distortion of other surrounding linac systems in adjacent rooms can be solved. The solution using low-field MRI systems will not be part of this chapter; the design of such a system can be found in Fallone et al. (2008, 2009).

11.2 Design Magnetic Resonance Linac

The design of the magnetic resonance linac (MRL) has been described in Lagendijk et al. 2002 and 2008. To assure the diagnostic image quality of the system, we stayed with the basic design of a closed bore 1.5 T Philips Achieva system. The active shielding coils of the superconducting magnet of this system were modified such that at a distance of 80–100 cm from the housing of the cryostat, an area was created without magnetic field present in it. In this toroid-shaped area, circumferential to the magnet, the magnetic field strength is below 1 mT. By positioning the sensitive gun section of the linac in this void, the magnetic field at the linac is so small that the coupling between the systems is minimal, thus guaranteeing the functioning of both the linac and the MRI.

The positioning of the linac at the central plane of the MRI implies that the beam has to pass the MRI. We modified the magnet such that a transparent and homogeneous window is created in the cryostat to let the beam pass. All coils and system heterogeneities are moved out of the central plane creating a gap with a 24-cm projection in the isocenter. Also, the gradient coils are split to allow undisturbed beam passage. The total mass in this window is minimized to reduce the beam absorption and the photon scatter towards the patient. The remaining total thickness of the present system is equivalent to about 10 cm of aluminum. In the final clinical system, this will be reduced to about 5 cm of homogeneous aluminum. Also, due to the geometry chosen the isocenter target distance is about 1.4 m, requiring a special high-output linac and dedicated MLC. The RF coils and table support of the MRL will be designed transparent for the megavoltage beam. Figure 11.1 gives an artist impression of the system design.

11.3 Status

In March 2009, the mark 0 version of the system was installed in one of our treatment rooms. This version consists of the modified magnet, the split gradient coil, and an Elekta compact accelerator operating at 6 MV. The linac lacks gantry rotation and has been positioned stationary on a wooden table. The linac fires horizontally toward the isocenter in the MRI (Figure 11.2). The magnet and gradient coil have diagnostic specifications, and because the electronics and peripherals are from a standard Philips Achieva system, the imaging proved to be of a diagnostic quality (Figure 11.3). To prevent RF distortion of the MRI, the linac has been placed outside a modified Faraday cage. The impact of the magnetic field on the neighboring accelerators was

countered (Kok et al. 2009). The simultaneous operation of the linac and the MRI was demonstrated with the imaging of a piece of a pork chop (Figure 11.4). No image deterioration was detected when switching the beam on (Raaymakers et al. 2009). Image subtraction showed no differences between the beam on and the beam off situation, proving the simultaneous functioning of the MRL and the possibility to have real-time image guidance.

11.4 Dosimetry and Treatment Planning

The photon beam in the MRL cannot be affected by the magnetic field. Secondary electrons released by the photon beam are experiencing the Lorentz force and this causes an altered dose distribution (Raaymakers et al. 2004). The build-up distance of the beam is slightly reduced while the whole beam experienced a 0.6-mm offset in the direction of the Lorentz force (Raaymakers et al. 2004). The difference in dose distributions is most prominent at tissue–air interfaces. Without magnetic field, the electrons scatter away from the tissue; with a magnetic field, the electrons are forced back into the tissue by the Lorentz force resulting in an increased dose at the exit side of the beam (Raaijmakers et al. 2005), which is the electron return effect (ERE) (see also Figure 11.5). The ERE is dependent on the electron energies and the strength of the magnetic field, making it also dependent on the width of the radiation field (Raaijmakers et al. 2008; Kirkby et al. 2008); see Figure 11.6 for an illustration. The magnitude of this effect is also dependent on the orientation of the exit side with respect to the radiation beam and the magnetic field (Raaijmakers et al. 2007a).

The impact of the magnetic field on the dose distribution can be well described by using Monte Carlo dose computations. We used the Geant4 code developed at CERN (Agostinelli et al. 2003).

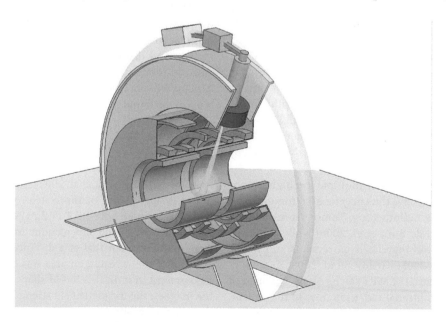

FIGURE 11.1 Artistic impression of the 1.5 T MRI accelerator system. The 6 MV accelerator is mounted in a ring around the MRI with the most critical accelerator components located in the low-magnetic-field toroid (indicated in transparent). The superconduction coils are repositioned in order to create this low field toroid as well as to create a beam portal. Also the gradient coil is split to allow beam passage.

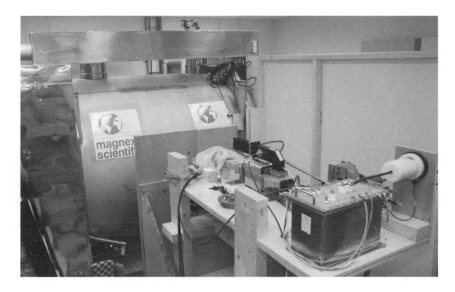

FIGURE 11.2 Photograph of the prototype MRI accelerator. The accelerator is positioned on the wooden stand; the magnet is shown behind it. Also, the copper RF cage at the service side of the magnet is shown.

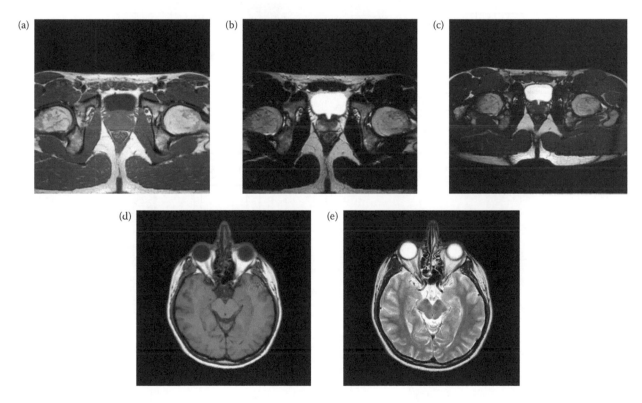

FIGURE 11.3 MRI data of volunteers from the 1.5 T MRI accelerator. (a) shows a T1 weighted Spin Echo of the pelvis with TR/TE of 663/15 ms and an acquired voxel size of $1 \times 1.25 \times 3$ mm^3. It clearly shows the prostate, rectum and bladder. (b) shows the same anatomy by means of a T2 weighted Turbo Spin Echo with TR/TE of 5989/20 ms and an acquired voxel size of $1 \times 1.5 \times 3$ mm^3, in (c) a balanced Steady State Free Precession scan with TR/TE 6.6/3.3 ms and an acquired voxel size of $1.77 \times 1.33 \times 2.03$ mm^3 is shown. Another example is from the brain, in (d) a T1 weighted Spin Echo with TR/TE 596/15 ms and an acquired voxel size of $1.02 \times 1.27 \times 5.0$ mm^3, while (e) shows the same brain by means of a T2 weighted Turbo Spin Echo TR/TE 4892/110 ms and acquired voxel size of $0.65 \times 0.84 \times 5.0$ mm^3.

The impact of the magnetic field is compensated in first order by choosing opposing beams or multiple beam angles (Raaijmakers et al. 2005). By including the impact of the magnetic field in the inverse optimization process the remaining impact can be countered (Raaijmakers et al. 2007b). It was shown that by using a multibeam setup in combination with inverse optimization resulted in the same dose distributions as for the case without magnetic field. The ERE is present at any density change in the

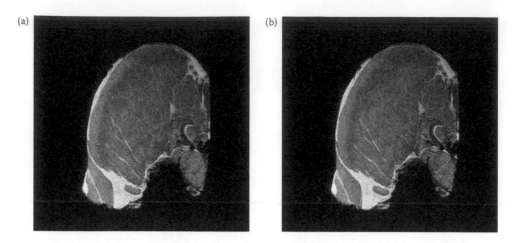

FIGURE 11.4 MRI of a piece of pork chop without (a) and with (b) the radiation beam on. A T2 weighted Spin Echo sequence using a C1 surface coil (Philips, Best, the Netherlands) with TR/TE 592/100 ms and acquired voxel size of $0.53 \times 0.54 \times 5.0$ mm^3. The images with and without radiation beam on are identical, showing the proof of concept of simultaneous MR imaging and irradiation.

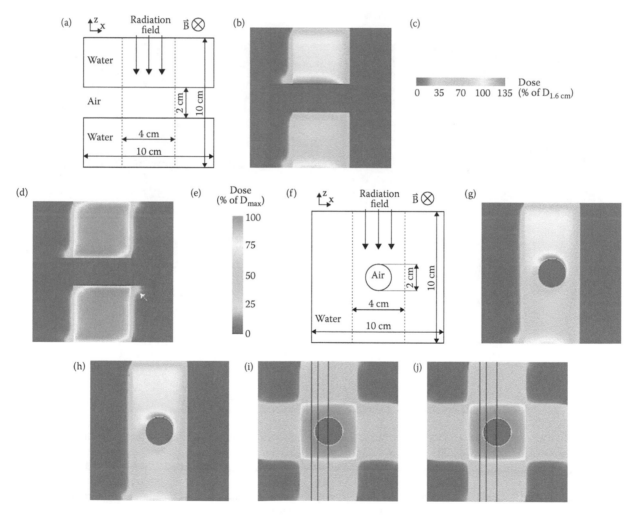

FIGURE 11.5 **(See color insert.)** The impact of a 1.5 T transverse magnetic field on the dose distribution at a tissue–air interface. In (a), the setup of a layered water–air–water phantom is shown. In (b), the dose distribution from a single beam with the large dose increase at the interfaces is shown and in (c), the dose distribution from opposing beams is shown where the dose distribution in first order is homogeneous again. In (d), the setup for a cylinder of air in a homogeneous water phantom is shown. In (e), the ERE effect for this geometry is shown and (f) shows that 4 beams are required to restore the homogeneity for this geometry. (Adapted from Raaijmakers, A.J.E., Raaymakers, B.W., Lagendijk, J.J.W., *Phys. Med. Biol.*, 50:1363–1376, 2005.)

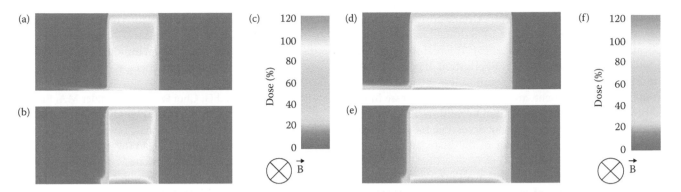

FIGURE 11.6 **(See color insert.)** The impact of the magnetic field on the dose distribution is dependent on the magnetic field strength and the width of the radiation field. In (a) and (b), the dose distribution for a 5 × 5 cm² field is shown at 0.2 T and at 1.5 T. At 1.5 T, the ERE is much more pronounced than for 0.2 T although the same number of electrons are returning to the phantom. However, at 0.2 T, these electrons have large radii and enter the phantom again outside the radiation field. This is clearer in (c) and (d), where the dose distribution for a 10 × 10 cm² at 0.2 T and 1.5 T are shown. Now the amplitude of the ERE is approximately the same, but at 0.2 T the out-of-field dose deposition by the ERE is clearly seen on the left hand side of (c). (Adapted from Raaijmakers A.J.E., Raaymakers B.W., Lagendijk J.J.W., *Phys. Med. Biol.* 53(4): 909–23, 2008.

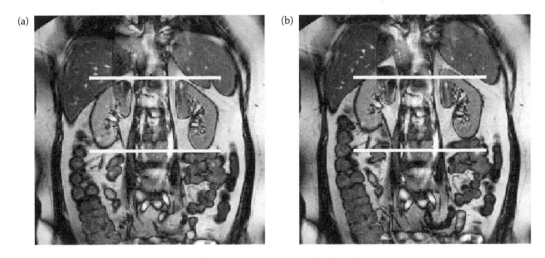

FIGURE 11.7 Two stills from a cine MRI showing kidney motion during free breathing. A dynamic series of 34 images using a coronal single slice balanced steady state free precession (bSSFP) sequence with an image acquisition time of about 450 ms and an in plane resolution of about 1.3 mm was used. (a) Shows the still at end-inhale state and (b) shows the end-exhale state. The white lines are added for comparison purposes. The tumor is visible as the caudal small white lesion in the right kidney.

tissue; the tissue–lung interface displays a strong effect (Kirkby et al. 2008). This effect may not always be countered by opposing beams because it occurs at the lung surface. However, for almost all clinical applications, the dose increase induced stays well below the clinical relevant intensities in multiple beam arrangements (Raaijmakers et al. 2007b).

Absolute dosimetry within a magnetic field is possible by describing and correcting the impact of the magnetic field on standard ionization chambers (Meijsing et al. 2009).

An appealing feature of a radiotherapy system with integrated MRI functionality is the possibility of 3D dosimetry using Fricke (Gore et al. 1984) or polymer gels (Vergote et al. 2003). These phantoms have dose-dependent MRI properties and can be read simultaneously with dose delivery, potentially facilitating 4D dose quantification. Even more appealing is the possibility of application of beam visualization directly in the patient. We are investigating if, for very specific MRI sequences, the secondary electrons released by the photon beam can alter the MRI images.

11.5 Clinical Impact

In radiotherapy, significant margins around the GTV and CTV are being used to assure that the tumors stayed in the field during daily fractionation (ICRU 50). The geometrical fact that the biggest volume is always in the outer shell (Verellen et al. 2007) implies that large volumes of normal tissue are being irradiated and that the tumor dose is limited by normal tissue complications. As a result, radiotherapy is often combined with surgery to remove the central tumor mass. Present day radiotherapy is at its best killing small tumor infiltrations in normal tissue while sparing the tissue. At present, extreme position techniques allow stereotactic irradiation also for a limited number of body applications (van der Pool et al. 2010). The real-time images at the MRL will greatly increase the application of body stereotactic techniques and will directly result in less integral dose, less normal tissue involvement, and thus a further increase in realized tumor dose, given the potential to minimize or avoid surgery also for locations presently not accessible for radiotherapy, such as renal tumors and tumors of the gastrointestinal track (Kerkhof et al. 2010) (Figure 11.7).

It is our strong belief that the MRL has the potential to become the next generation radiotherapy standard and that this system allows opening a range of new radiotherapy applications, thus minimizing the need for surgery.

References

Agostinelli S et al., Geant4—A simulation toolkit *Nucl. Instrum. Methods Phys. Res. A,* 505:250–303, 2003.

Barentsz JO, Fütterer JJ, and Takahashi S, Use of ultra-small superparamagnetic iron oxide in lymph node MR imaging in prostate cancer patients. *Eur J Radiol.,* 63(3):369–72, 2007.

Fallone BG, Murray B, Rathee S, Stanescu T, Steciw S, Vidakovic S, Blosser E, and Tymofichuk D. First MR images obtained during megavoltage photon irradiation from a prototype integrated linac-MR system. *Med. Phys.* 36(6):2084–2088, 2008.

Fallone BG. Real-Time MR-Guided Radiotherapy: Integration of a Low-Field MR System *Med. Phys.* 36(6):2774–2775, 2009.

Fütterer JJ, Barentsz JO, and Heijmink SW, Value of 3-T magnetic resonance imaging in local staging of prostate cancer, *Top Magn Reson Imaging.* 19(6):285–9, 2008.

Ghilezan MJ, Jaffray DA, Siewerdsen JH, Van Herk M, Shetty A, Sharpe MB, Zafar Jafri S, Vicini FA, Matter RC, Brabbins DS, and Martinez AA. Prostate gland motion assessed with cine-magnetic resonance imaging (cine-MRI). *Int. J. Radiat. Oncol. Biol. Phys.* 62(2):406–17, 2005.

Groenendaal G, Moman MR, Korporaal JG, van Diest PJ, van Vulpen M, Philippens ME, and van der Heide UA. Validation of functional imaging with pathology for tumor delineation in the prostate. *Radiother. Oncol.,* 84(2):145–50, 2010.

Gore JC, Kang YS, and Schulz RJ Measurement of radiation dose distributions by nuclear magnetic resonance (NMR) imaging. *Phys. Med. Biol.* 29(10):1189–97, 1984.

Jackson AS, Reinsberg SA, Sohaib SA, Charles-Edwards EM, Jhavar S, Christmas TJ, Thompson AC, Bailey MJ, Corbishley CM, Fisher C, Leach MO, and Dearnaley DP, Dynamic contrast-enhanced MRI for prostate cancer localization, *J. Radiol.* 2009; 82(974):148–56.

Kerkhof EM, van der Put RW, Raaymakers BW, van der Heide UA, Jürgenliemk-Schulz IM, and Lagendijk JJW Intrafraction motion in patients with cervical cancer: The benefit of soft tissue registration using MRI. *Radiother. Oncol.,* 93(1):115–21, 2009.

Kerkhof EM, Raaymakers BW, van Vulpen M, Zonnenberg BA, Bosch JLR, van Moorselaar RJ Lagendijk JJW, A new concept for noninvasive renal tumour ablation using real-time MRI guided radiation therapy. *BJU Int.,* 107(1):63–8, 2011.

Kirilova A, Lockwood G, Choi P, Bana N, Haider MA, Brock KK, Eccles C, and Dawson LA Three-dimensional motion of liver tumors using cine-magnetic resonance imaging. *Int. J. Radiat. Oncol. Biol. Phys.* 71(4):1189–95, 2008.

Kirkby C, Stanescu T, Rathee S, Carlone M, Murray B, and Fallone BG Patient dosimetry for hybrid MRI-radiotherapy systems. *Med. Phys.* 35(3):1019–27, 2008.

Klerkx WM, Mali WM, Peter Heintz A, de Kort GA, Takahara T, Peeters PH, Observer variation of magnetic resonance imaging and diffusion weighted imaging in pelvic lymph node detection. *Eur. J. Radiol.,* 78(1):71–4, 2009.

Klomp DW, Bitz AK, Heerschap A, Scheenen TW, Proton spectroscopic imaging of the human prostate at 7 T. *NMR Biomed.* 22(5):495–501, 2009.

Kok JGM, Raaymakers BW, Lagendijk JJW, Overweg J, de Graaff CH, Brown KJ Installation of the 1.5 T MRI accelerator next to clinical accelerators: impact of the fringe field. *Phys. Med. Biol.* 54(18):N409–15, 2009.

Korporaal JG, van den Berg CA, Groenendaal G, Moman MR, van Vulpen M, van der Heide UA, The use of probability maps to deal with the uncertainties in prostate cancer delineation, *Radiother. Oncol.* 94(2):168–72, 2010.

Kurhanewicz J, Vigneron DB, Advances in MR spectroscopy of the prostate, *Magn. Reson. Imaging Clin. N. Am.* 16(4):697–710, 2008.

Lagendijk, JJW, Raaymakers BW, Van der Heide UA, Topolnjak R, Dehnad H, Hofman P, Nederveen AJ, Shulz IM, Welleweerd J, Bakker CJG MRI guided radiotherapy: MRI as position verification system for IMRT, *Radiother. Oncol.* 64(S1):S75–S76, 2002.

Lagendijk JJW, Raaymakers BW, Raaijmakers AJE, Overweg J, Brown KJ, Kerkhof EM, van der Put RW, Hårdemark B, van Vulpen M, van der Heide UA MRI/linac integration, *Radiother. Oncol.* 86(1):25–9, 2008.

Meijsing I, Raaymakers BW, Raaijmakers AJ, Kok JG, Hogeweg L, Liu B, Lagendijk JJ, Dosimetry for the MRI accelerator: the impact of a magnetic field on the response of a Farmer NE2571 ionization chamber, *Phys. Med. Biol.* 54(10):2993–3002, 2009.

Moman MR, van den Berg CA, Boeken Kruger AE, Battermann JJ, Moerland MA, van der Heide UA, van Vulpen M, Focal Salvage Guided by T(2)-Weighted and Dynamic Contrast-Enhanced Magnetic Resonance Imaging for Prostate Cancer Recurrences, *Int. J. Radiat. Oncol. Biol. Phys.* 76(3):741–746, 2010.

Raaijmakers AJE, Raaymakers BW, Lagendijk JJW Integrating a MRI scanner with a 6 MV radiotherapy accelerator: dose increase at tissue–air interfaces in a lateral magnetic field due to returning electrons *Phys. Med. Biol.* 50(7):1363–76, 2005.

Raaijmakers AJE, Raaymakers BW, Lagendijk JJW Magnetic-field-induced dose effects in MR-guided radiotherapy systems: dependence on the magnetic field strength. *Phys. Med. Biol.* 53(4): 909–23, 2008.

Raaijmakers AJE, Raaymakers BW, van der Meer S, Lagendijk JJW Integrating a MRI scanner with a 6 MV radiotherapy accelerator: impact of the surface orientation on the entrance and exit dose due to the transverse magnetic field. *Phys. Med. Biol.* 52(4):929–39, 2007a.

Raaijmakers AJE, Hardemark B, Raaymakers BW, Raaijmakers CP, Lagendijk JJW Dose optimization for the MRI-accelerator: IMRT in the presence of a magnetic field. *Phys. Med. Biol.* 52(23):7045–54, 2007b.

Raaymakers BW, Raaijmakers AJE, Kotte ANTJ, Jette D, Lagendijk JJW Integrating a MRI scanner with a 6 MV radiotherapy accelerator: dose deposition in a transverse magnetic field. *Phys. Med. Biol.* 49(17):4109–18, 2004.

Raaymakers BW, Lagendijk JJW, Overweg J, Kok JGM, Raaijmakers AJE, Kerkhof EM, van der Put RW, Meijsing I, Crijns SPM, Benedosso F, van Vulpen M, de Graaff CH, Allen J, Brown KJ Integrating a 1.5 T MRI scanner with a 6 MV accelerator: proof of concept. *Phys. Med. Biol.* 54(12):N229–37, 2009.

van de Bunt L, Jürgenliemk-Schulz IM, de Kort GA, Roesink JM, Tersteeg RJ, van der Heide UA, Motion and deformation of the target volumes during IMRT for cervical cancer: what margins do we need? *Radiother. Oncol.* 88(2):233–40, 2008.

van der Pool AE, Méndez Romero A, Wunderink W, Heijmen BJ, Levendag PC, Verhoef C, Ijzermans JN, Stereotactic body radiation therapy for colorectal liver metastases, *Br. J. Surg.* 97(3):377–82, 2010.

Verellen D, Ridder MD, Linthout N, Tournel K, Soete G, Storme G, Innovations in image-guided radiotherapy, *Nat. Rev. Cancer.* 7(12):949–60, 2007.

Vergote K, De Deene Y, Claus F, De Gersem W, Van Duyse B, Paelinck L, Achten E, De Neve W, De Wagter C Application of monomer/polymer gel dosimetry to study the effects of tissue inhomogeneities on intensity-modulated radiation therapy (IMRT) dose distributions. *Radiother. Oncol.* 67(1):119–28, 2003.

The ViewRay™ System

12.1 Introduction..115
12.2 Historical Perspective..115
    Cobalt and Conformal Therapy • Technical Developments • IMRT Changes the Game
    • Patient and Tumor Positioning • IGRT Begins for Radiotherapy: CT "Snapshots"
    • MRI versus CT Imaging for Intrafraction Organ Motion
12.3 Gamma-Ray IMRT..119
    Gamma-Ray IMRT Treatment Plans
12.4 Radiotherapy and MRI..121
    Magnetic Field Selection • The ViewRay™ System
References.. 126

Daniel A. Low

Richard Stark

James F. Dempsey

12.1 Introduction

This chapter describes the merging of intensity modulated radiation therapy (IMRT) and magnetic resonance imaging (MRI) in a novel device being built by ViewRay, Inc. The ViewRay™ System is composed of an open cylindrical MRI system coupled with a cobalt-based IMRT system. The advantage of using cobalt is that the IMRT unit does not require the high radiofrequency energy density needed for traditional radiation therapy devices. In the era of IMRT, beam energy selection does not impact treatment plan quality as significantly as it did during the era of two- and three-dimensional conformal radiation therapy; hence, cobalt-based IMRT can provide dose distributions nearly equivalent to linear accelerator-produced beams. The prospect of having real-time MRI for positioning, motion monitoring and gating, and eventually functional imaging is exciting, and the ViewRay™ System is the first commercial system that will allow this.

12.2 Historical Perspective

Clinical cobalt therapy units and MV linear accelerators were introduced nearly simultaneously in the early 1950s. The first two clinical cobalt therapy units were installed in October 1951 in Saskatoon and London, Ontario (Litt 2000). The first MV linear accelerator installed solely for clinical use was at Hammersmith Hospital, London in June 1952 (Bernier et al. 2004). In August 1953, the first patient was treated with this machine. The deeply penetrating ionizing photon beams quickly became the mainstay of radiation therapy, allowing the widespread noninvasive treatment of deep-seated tumors. An additional advantage of these photon beams was based on the fact that photons deliver most of their dose through the interactions of highly energetic

scattered electrons. The dose delivery properties of the photon/scattered electron system leave the patient's skin surface with a considerably lower dose than inside the body. This allowed aggressive doses to be used internally while sparing the skin from severe radiation damage. Consequently, this feature was called skin sparing and played an important role in the clinical utility of linear-accelerator and cobalt-beam therapies.

As expertise in the use of these techniques improved, and diagnostic technologies allowed cancers to be detected earlier, the efficacy of radiation therapy improved, and the role of x-ray therapy slowly changed from a mainly palliative (or pain relieving) therapy to a definitive curative therapy. The invention of these devices and the realization of the potential to refine our use of external beams to elegantly shape and control dose distributions to cure cancer while sparing the patient from undesirable and possibly life-threatening side effects set radiotherapy upon an unparalleled course of development that would eventually lead to the technical innovations of today.

12.2.1 Cobalt and Conformal Therapy

Despite similarities, cobalt units and linear accelerators were always viewed as rival technologies for external beam radiotherapy, and this rivalry would result in the eventual demise of one, at least in the United States and Western Europe. The cobalt unit relied on the formidable technology of a high-neutron-flux nuclear-fission reactor to produce the radioactive ^{60}Co isotope that produces 1.17 and 1.33 MeV photons; however, the device itself was quite simple and was not technically improved significantly over time. Of course, the simplicity of the cobalt unit was a cause for some of its appeal; the cobalt units were very safe, reliable, precise, and required little maintenance and technical expertise to run. Early on, this allowed cobalt therapy to become the most widespread form of external beam therapy. In fact, cobalt units dominated the radiotherapy external beam market from 1951 to 1984. The linear accelerator was a more technically intensive device, bringing the state-of-the-art accelerator technology out of the nuclear physics research lab and to the clinic. Accelerating high currents of electrons to energies between 4 and 25 MeV to produce beams of bremsstrahlung photons or scattered electrons, the linear accelerator was a much more versatile machine that allowed more penetrating beams with sharper penumbrae and higher dose rates.

Before the development of IMRT, radiation therapy was delivered using multiple custom-shaped intersecting portals. The portal shapes were designed to conform to the shape of the tumor or the region where the tumor was suspected to be. The intersection of these beams formed a region that had greatly increased dose relative to the surrounding regions, allowing treatment plans to be designed that avoided overdosing critical structures while treating the tumor to sufficiently high doses. Still, there were many cases when the radiation beams were aimed toward each other at 180° angles (parallel, opposed). The dose falloff, caused by beam attenuation and the inverse-square fluence reduction effect, was partially cancelled by the opposed beams.

While the intersecting or opposed beams delivered the prescription dose to the tumor, the magnitude of the dose in the surrounding tissue was a function of the tumor depth (often a function of the obesity state of the patient) and the beam energy. The greater the tumor depth (patient thickness), the greater was the undesirable hot spot in the normal tissues. The greater the linear accelerator-generated photon beam energy, the lower the hot spot. Given that much of this development occurred before the advent of three-dimensional conformal therapy and the subsequent quantitative evaluation of normal organ doses, the lower the hot spot, the better. Linear accelerators that provided multiple beam energies, all greater in penetration than what cobalt could provide, were desired by clinics that wanted to treat these deep-seated tumors.

In addition to the high-energy photon beams, linear accelerators are capable of producing high-energy scattered electron beams. These are uniquely suited to treating superficial tumors or suspected tumor bearing regions. Before the discovery of IMRT, electron beams provided a superior method for treating breast and head-and-neck cancer. This gave the linear accelerator a competitive advantage over the cobalt unit for almost two decades.

12.2.2 Technical Developments

As the linear accelerator became more reliable, the benefits of having more penetrating photon beams coupled with the addition of electron beams was seen as strong enough impetus to replace the existing cobalt units. Cobalt therapy did not die away without some protests, and the essence of this debate was captured in a famous paper in 1986 by Laughlin, Mohan, and Kutcher (Laughlin et al. 1986), which explained the pros and cons of cobalt units against linear accelerators. This was accompanied by an editorial from Suit (Suit 1986) that pleaded for the continuance and further technical development of cobalt

units. The pros of cobalt units and linear accelerators have been already been listed. The cons of cobalt units were seen as less penetrating photons, larger penumbra due to the projection of the cobalt source size, greater surface doses for large fields due to low energy contamination electrons, and mandatory regulatory oversight. Interestingly, the cons for linear accelerators increased with their increasing energy (and hence their difference from a low-energy cobalt beam) and were seen to be increased penumbra due to secondary electron transport, increased dose to bone (due to increased dose due to pair production), and most importantly, the production of photoneutrons at acceleration potentials over 10 MV.

It is clear that in the era before IMRT, linear accelerators held definite advantages over cobalt therapy when we sculpted dose distributions using a handful of manually weighted beams. However, some concerns that the benefits of linear accelerators were not sufficient to justify their cost and complexity were voiced, and perhaps some unwarranted enthusiasm or even hyperbole was contained in the overall preference of linear accelerators over cobalt (Suit 1986).

Equipping a clinic with a combination of cobalt units and linear accelerators could provide the desired clinical functionality at far reduced cost than eliminating the cobalt units. The commercial vendors of linear accelerators had a great incentive to compete against cobalt units. Their purchase price was greater due to their increased complexity, and this complexity led to increased service contract revenues. The linear accelerator manufacturers also had to compete with each other, and starting in the 1980s, began a rapid and, to this date, unending development of features that the cobalt manufacturers did not duplicate.

These technical improvements included computer-controlled asymmetric collimator jaws, multileaf collimators, and patient couches that were introduced for linear accelerators at this time (nothing would have prevented their use in cobalt units). Cobalt unit manufacturers retreated into less technically developed foreign markets. Thus, the value of cobalt therapy would lie dormant and unimproved for several decades. It is very interesting that the very last sentence of the work by Laughlin et al. was a qualification of their conclusions that would foreshadow the reinvestigation of cobalt for advanced radiotherapy: "Another consideration for the future is that the implementation of multileaf collimation and dynamic therapy, which is equivalent to the use of many fields, may further reduce the need for higher energies" (Laughlin et al. 1986). Of course, this discussion occurred long before the advent of large-scale optimization and IMRT in radiation therapy. According to experts in the development and history of IMRT (Webb 2001a), the concept of multileaf collimator (MLC)-based IMRT would not be articulated until two years later in 1988 by Kallman et al. (Kallman et al. 1988) and would not reach widespread clinical implementation for almost another decade.

12.2.3 IMRT Changes the Game

The introduction of megavoltage photon-beam IMRT achieved using MLC delivery represents a significant advance with great potential for the discipline of radiation therapy (Webb 2001a, b; IMRTCWG 2001). While IMRT may be delivered using many mechanisms including physical attenuators, after-market binary collimators, and accelerators designed to allow scanned or binary-modulated beams (Webb 2001a; IMRTCWG 2001), the MLC delivery methodology has become prevalent due to its integration into existing delivery systems. This development has allowed the widespread clinical implementation of very complex nonconvex dose distributions. Not surprisingly, this complex but promising treatment modality has rapidly proliferated in both academic and community practice settings. In principle, the complex dose distributions available with IMRT will allow escalation of the ionizing radiation dose that can be delivered while lowering or maintaining the undesirable side effects caused by radiotherapy. Indeed, evidence of the advantageous employment of this IMRT MLC delivery to reduce treatment-related toxicity has already been published demonstrating that IMRT can spare salivary gland function (Chao et al. 2001a, b; Eisbruch et al. 1999) in head-and-neck radiotherapy and reduce bowel toxicity in prostate radiotherapy (Zelefsky et al. 2000; Zelefsky et al. 2001).

MLC-based IMRT represents a significant departure from previous radiotherapy practice in that the tumoricidal high-dose regions and tissue sparing low-dose regions are constructed via the superposition of many steep-gradient penumbrae from small MLC-shaped fields. This is achieved by relying on the solution of a unique large-scale optimization problem to determine the radiation fluence distributions for each individual patient (Shepard et al. 1999). In the past, when the value of cobalt therapy was weighed against linear accelerator-based therapy treatment, fields were manually developed without the benefit of these powerful optimization techniques. As IMRT developed, the use of higher energy photon beams and electron beams was largely abandoned by the community. The TomoTherapy HiArt System® is an excellent example of a versatile IMRT machine, which utilizes only a relatively low-energy 5.25 MV photon beam (Thomas et al. 2005). Employment of a low-energy beam is partly due to the increased concern over neutron production for the increased beam-on times required by IMRT and the greater size of accelerator for higher-energy beams. Mostly, it is due to the fact that low MV photon-beam IMRT can produce treatment plans of excellent quality for practically any site of cancer treatment and more penetrating beams are simply not needed. While the advances of IMRT have been significant, the development of radiation therapy is ongoing. This is due to the fact that the technology of radiation therapy is out of balance; while we have an amazing facility to optimize and calculate accurate doses to static objects, we have very little ability to account for the fact that the patient is (hopefully!) a dynamic living, breathing, and changing being (which, by the way, is a very good thing!).

12.2.4 Patient and Tumor Positioning

IMRT represents a culmination of decades of improving 3D dose calculations and large-scale optimization to the point that we have achieved a high degree of accuracy and precision for static objects. However, there is a fundamental flaw in our currently accepted paradigm for dose modeling. The problem lies with the fact that patients are essentially squishy articulated sacks of tissue that we cannot and will not perfectly reposition for fractioned radiotherapy. Despite this fact, the delivery of radiation therapy is traditionally planned on a static model of radiotherapy targets and critical structures. The real problem lies in the fact that outside of the cranium (i.e., excluding the treatment of CNS disease using Stereotactic radiotherapy), radiation therapy needs to be fractionated to be effective, that is, it must be delivered in single 1.8–2.2 Gy fractions or double 1.2–1.5 Gy fractions daily, and it is traditionally delivered during the work week (Monday through Friday), taking 7–8 weeks to deliver a curative dose of 70–72 Gy at 2.0 or 1.8 Gy. This daily fractionation requires the patient and all of their internal organs to be repositioned exactly for accurate dose delivery. This raises an extremely important question for radiation therapy: "Of what use is all of the elegant dose computation and IMRT optimization we have developed if the targets and critical structures move around during the actual therapy?" It is well known that exactly repositioning the patient is not possible due to several factors, including the inability to reproduce the *patient setup*, that is, the gross tissue geometry and alignment of the patients boney anatomy; *physiological changes* in the patient, such as weight loss or tumor growth and shrinkage; and *organ motion* in the patients including, but not limited to, breathing motion, cardiac motion, rectal distension, peristalsis, bladder filling, and voluntary muscular motion.

In order to attempt to account for organ motions, the concept of margins and planning target volumes (PTV) was developed to irradiate a volume that was expected to contain the target during most of the irradiation (ICRU 1993; ICRU 1999). An excellent critical review of organ motion studies was published by Langen and Jones in 2001 (Langen and Jones 2001). They summarized the existing literature up to that point regarding the two most prevalent types of organ motion: patient setup errors and dynamic organ movement. While significant physiological changes in the patient do occur, for example, tumor shrinkage in head-and-neck cancer, they have not been well studied. The organ motion studies were further subdivided into interfraction (changes between daily deliveries) and intrafraction (changes during a given daily delivery) organ motion, with the acknowledgement that the two could not be explicitly separated, that is, intrafraction motions obviously confound the clean observation of interfraction motions. Published data on interfraction motion of gynecological tumors, prostate, bladder, and rectum were reviewed as well as published data on the intrafraction movement of the liver, diaphragm, kidneys, pancreas, lung tumors, and prostate. Published methods for managing interfraction and intrafraction organ motion in radiation therapy were also reviewed. Sixty-one peer-reviewed publications were summarized in this review, spanning the two decades before publication. The fact that both inter and intrafraction organ motions can have a significant effect on

TABLE 12.1 Summary of Organ Motion Studies

Organ/Tumor	No. of Studies	No. of Patients	Motion Range [mm]
Interfraction motion			
Bladder	7	11–30	27 A.P.
			4% Vol. loss per week
			40–80% Vol. change
Gynecological tumors	1	29	<7 Sup.
			<4 Pos.
Prostate	18	6–55	5.3–20.0 A.P.
			1.7–9.9 S.I.
			2.0–8.8 Lat.
Rectum	5	11–30	17–76 Diameter change
			6%/Week vol. decrease
Seminal vesicles	5	6–50	1.5–22.0 A.P.
			0.35–14.0 S.I.
			0.3–5.5 Lat.
Intrafraction motion			
Diaphragm	6	5–30	5–40 Normal breathing
			25–80 Deep breathing
Kidneys	6	8–100	2–40 Normal breathing
			4–86 Deep breathing
Liver	5	9–50	7–38 Normal breathing
			10–103 Deep breathing
Lung tumors	2	20	5–22 A.P.
			0–16 Lat.
			1.3–6.5 S.I.
Pancreas	2	36–50	10–30 Normal breathing
			20–80 Deep breathing
Prostate	3	55	No Motion in EPID
			0–15 Transient motion with Ciné MRI

radiation therapy dosimetry can be easily seen in Table 12.1, where displacements between 0.5 and 4.0 cm have been commonly observed in studies involving less than 50 patients. The mean displacements for many observations of an organ motion can be small, but even an infrequent yet large displacement can significantly alter the biologically effective dose received by a patient, as it is well accepted that the correct dose per fraction must be maintained to effect tumor control (Hall and Giaccia 2006).

Over the last decade, the field of radiation oncology has been slowly turning its attention to dealing with all forms of organ motion including the most elusive and possibly most significant; intrafraction motion. In a more focused review of intrafraction organ motion recently published by Goitein (2004, p.5), the importance of dealing with organ motion-related dosimetry errors was concisely stated: "… it is incontestable that unacceptably, or at least undesirably, large motions can occur in some patients, …" It was further explained by Goitein (2004, p.2) that the problem of organ motions has always been a concern in radiation therapy: "We have known that patients move and breathe and that their hearts beat and their intestines wriggle since radiation was first used in cancer therapy. In not-so-distant decades, our solution was simply to watch all that motion on the simulator's fluoroscope and then set the field edge wires wide enough that the target (never mind that we could not see it) stayed within the field." The real point is not that organ motion is a recent discovery, but rather that imaging, computational, and technological advancements have been made, where addressing organ motion problems becomes possible.

12.2.5 IGRT Begins for Radiotherapy: CT "Snapshots"

In an attempt to partially address the limitations imposed on radiation therapy by patient setup errors, physiological changes, and organ motion throughout the protracted weeks of radiation therapy, the three major linear accelerator manufacturers (Varian, Siemens, and Elekta) along with a company that provides a linear accelerator-based helical tomotherapy IMRT unit (Tomotherapy) have produced devices that integrate an imaging system capable of acquiring a volumetric CT "snapshot" before and after each daily delivery of radiation. This new combination of a state-of-the-art radiation therapy unit with radiology imaging equipment has been termed image-guided radiation therapy (IGRT), or preferably image guided IMRT (IGIMRT) (25–34). This advancement is a significant step toward placing the technological focus of radiation therapy back into balance. This existing technology has the potential for removing patient setup errors and slow physiological changes that occur over the extended course of radiotherapy. Currently, the devices are only being used to shift the gross patient position, but the development of daily dose computation and daily IMRT treatment plan reoptimization is just around the corner. Although this technology is a great improvement over the previous methodology, it is not without some significant limitations.

First, the speed at which helical or cone-beam CT can be performed is fairly slow compared with the time scales involved in intrafraction organ motion, and such motion is challenging to account for (Langen and Jones 2001; Goitein 2004). Secondly, CT imaging adds to the ionizing radiation dose delivered to the patient. It is well known that secondary carcinogenesis can occur in regions of low-to-moderate dose and the whole body dose will be increased by the application of many CT image studies. For a single CT snapshot on a daily basis, the additional dose may be acceptable. However, to achieve real-time imaging, hundreds of images per fraction would have to be acquired making the dose unacceptably high; for example, even at a very low CT dose of 0.5 cGy per CT volume [currently available onboard CT doses range from 2 to 15 cGy depending on the site and require more than 1 minute of acquisition time, with lower doses obtainable in the head and neck and higher doses required in the pelvis (Forrest et al. 2004; Letourneau et al. 2005; Pouliot et al. 2005)] one image per second for 5 minutes of beam-on-time would result in 150 cGy per day—a large therapeutic dose, which would produce acute tissue effects to the imaged volume (quite possibly lethal to the patient). Clearly, CT-based real-time volumetric imaging is not currently technically feasible and will require an approximate 100 fold or greater decrease in dose before it could even be considered a possibility.

12.2.6 MRI versus CT Imaging for Intrafraction Organ Motion

MRI is an attractive alternative to CT imaging. It involves no addition of dose to the patient, can image in arbitrary orientations without moving parts, is capable of extremely fast image acquisition, and provides soft tissue contrast that is unavailable with CT. Much like the development of cobalt therapy units and linear accelerators, CT imaging (Hounsfield 1973; Hounsfield 1976; Hounsfield 1977) and MRI (Lauterbur 1973; Lauterbur 1974; Lauterbur et al. 1976; Lauterbur et al, 1978; Mansfield 1977; Mansfield and Maudsley 1976) units were both demonstrated in the 1970s. Unlike the competition between cobalt units and linear accelerators, both CT imaging and MRI have thrived and undergone significant development, emerging today as mature imaging technologies. CT imaging was adopted as the gold standard for radiation therapy imaging early on due to its intrinsic spatial integrity. The integrity is a product of the physical processes of x-ray attenuation and the fixed geometry of the source and detectors. The utility of CT was also enhanced by the fact that the CT attenuation coefficients could be related to electron densities, the quantity required to compute radiation dose.

Despite the possibility of spatial distortions occurring in MRI, it is still very attractive as an imaging modality for radiotherapy because it has better soft tissue contrast than CT imaging (Khoo et al. 2004) and also has the ability to image physiological and metabolic information such as chemical tumor signals (Wu et al. 2004) or oxygenation levels (Taylor et al. 2001). The MRI artifacts that influence the spatial integrity of the data are related to undesired fluctuations in the magnetic field homogeneity and can be separated into two categories: (1) artifacts due to the scanner itself (e.g., field inhomogeneities intrinsic to the magnet design and induced eddy currents due to gradient switching), and (2) artifacts due to the imaging subject, that is, the intrinsic magnetic susceptibility of patients' tissues (Haacke 1999). Modern MRI units are carefully characterized (Tanneer et al. 2000) and employ reconstruction algorithms (Sutton et al. 2003) that can effectively eliminate the first type of artifacts. At high magnetic field strengths, in the range of 1.0–3.0 T, magnetic susceptibility of the patient tissues can produce significant distortions (which are proportional to field strength). These artifacts can often be eliminated by first acquiring several volumes of susceptibility imaging data and then using these data to develop corrections (Haacke 1999). Recently, many academic centers have started to employ MRI for radiation therapy treatment planning (Beavis et al. 1998; Chen et al. 2004; Krempien et al. 2002; Lee et al. 2003; Mah et al. 2002a, b; Mizowaki et al. 1996, 2000, 2001; Petersch et al. 2004; Prott et al. 2000; Shimizu et al. 2000; Steenbakkers et al. 2003). Rather than dealing with patient-related artifacts at high field (Chen et al. 2004; Mah et al. 2002a), many radiation therapy centers have employed low-field MRI units with 0.2–0.3 T for radiation therapy treatment planning, and a commercial 0.23 T MRI simulator has been available for years (Panorama 0.23T R/T, Philips Medical Systems). The magnetic field strength of these scanners is sufficiently low so that the patient-susceptibility spatial distortions are insignificant (Petersch et al. 2004).

12.3 Gamma-Ray IMRT

The combination of a linear accelerator and MRI unit, which is beyond the scope of this chapter, is challenging due to the interaction of the large radiofrequency fields required for electron acceleration and the quiet radiofrequency environment required for MRI acquisition. In 2001, at a little-noticed presentation in a biennial ESTRO physics meeting in Spain, Warrington and Adams (Warrington and Adams 2001) demonstrated that cobalt IMRT was indeed feasible and produced excellent results by commissioning their treatment planning system with beam data measured using a Theratronics 780C cobalt unit. Other than potential cost savings, there was no compelling motivation given to replace linear accelerator IMRT with cobalt IMRT.

A recent collaboration lead by Dr. Dempsey in the Department of Radiation Oncology and Dr. H. Edwin Romeijn in the Department of Industrial and Systems Engineering at the University of Florida lead to the development of very efficient and robust models of IMRT optimization (Romeijn et al. 2003, 2004). With the tools created by this collaboration and the perspective gained from running many trials on the in-house treatment planning system developed by Dr. Dempsey (the UFORT TPS), the group was able to investigate the question:

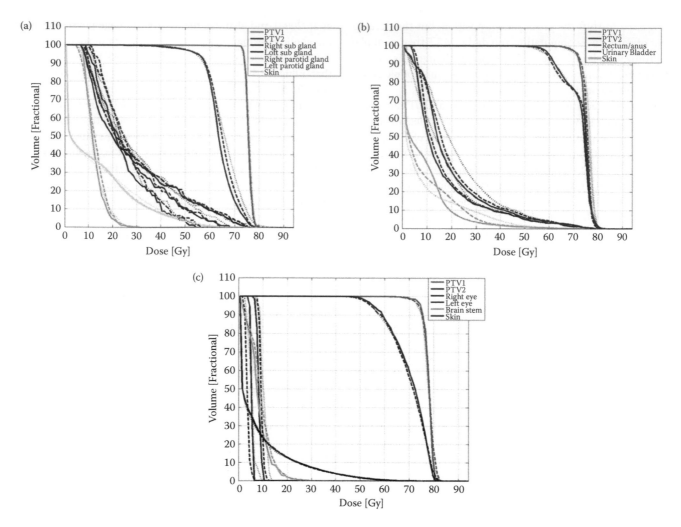

FIGURE 12.1 **(See color insert.)** Cobalt treatment plans for 5 (dotted), 9 (dashed) and 71 (solid) equidistant coplanar treatment plans. The plans are for the (a) head and neck, (b) prostate, and (c) central nervous system.

"How sharp do the beamlets need to be for effective IMRT?". Previous studies found that the clinical adequacy of head-and-neck IMRT treatment plans did not strongly depend on the number or orientation of coplanar 6-MV beams. A search for the limits of beam quality required to perform head-and-neck IMRT led to an investigation by Dr. Dempsey into the use of a cobalt unit beam that could be produced with a modern MLC. The treatment planning system was commissioned with a Theratronics 1000C cobalt beamlet model using measured radiochromic film data. The preliminary studies of the quality of a modernized form of cobalt therapy, termed γ-ray-based IMRT, were shown to be able to provide state-of-the-art IMRT dose distributions.

An open MRI unit and a γ-ray IMRT unit do not have the intrinsic conflicts of an MRI–linear accelerator combination. This opened the door for MRI to solve the organ motion problem and gave birth to the invention of the *ViewRay™ System*, which is presently being brought into the clinical marketplace by ViewRay, Inc. (Oakwood Village, Ohio).

12.3.1 Gamma-Ray IMRT Treatment Plans

An important consideration for the development of a unit that combines IMRT and MR guidance is the quality of cobalt-based IMRT. The poorer dose penetration of cobalt photons and the larger geometric penumbra due to the relatively large cobalt source might lead cobalt IMRT to be inferior to megavoltage photon IMRT. Fox et al. (2008) investigated this in 2008. They compared treatment plans conducted for a range of photon energies (6 MV, 18 MV, and cobalt), different numbers of static MLC delivered cobalt beams (5–71), and a helical tomotherapy cobalt beam geometry. The investigation used the convex fluence map optimization model, allowing the comparison of plan quality between different beam energies and configurations. They examined a number of relevant clinical sites, including head-and-neck, prostate, central nervous system, breast, and lung.

Figure 12.1 shows dose–volume histograms (DVHs) for the Head and Neck, prostate, and CNS clinical sites for cobalt treatment plans with 5, 9, and 71 equally spaced beams as prepared by

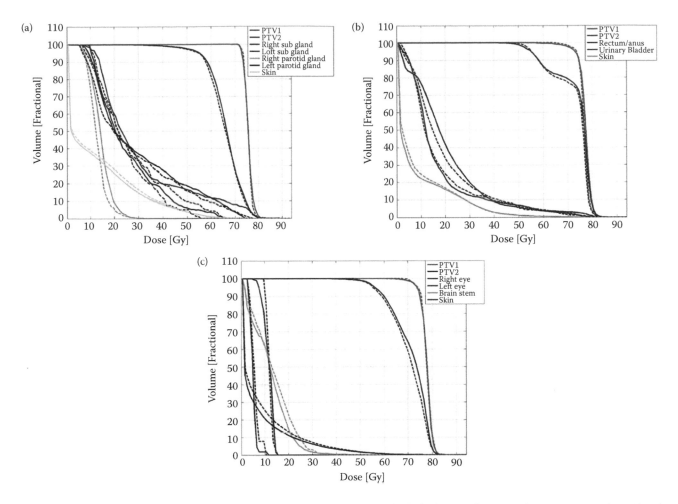

FIGURE 12.2 **(See color insert.)** IMRT treatment plans for 6 MV (solid) and cobalt (dashed) 7 equidistant coplanar treatment plans. The plans are for the (a) head and neck, (b) prostate, and (c) central nervous system.

Fox et al. (2008). The dose–volume histograms show that the differences between the treatment plans are minor when 9 or more fields are used. Figure 12.2 shows a comparison between 6 MV and cobalt IMRT for Head and Neck, prostate, and CNS clinical sites. The comparisons show that the two beam energies provide nearly identical treatment plans.

To further illustrate that cobalt IMRT can provide as conformal a radiation therapy as higher energy, head-and-neck IMRT treatment plans using 6 MV and cobalt are shown in Figures 12.3 and 12.4 for 7-field and 71-field IMRT (similar to a tomotherapy treatment).

12.4 Radiotherapy and MRI

12.4.1 Magnetic Field Selection

A charged particle moving in a vacuum at a velocity, $\vec{v}$, in the presence of a magnetic field, $\vec{B}$, experiences a Lorentz force given by $\vec{F} = q(\vec{v} \times \vec{B})$, which causes the particle to move in a circular motion. This force is not strong enough to significantly change the physics of the interactions of ionizing photons and electrons with matter; however, it can influence the overall transport of ionizing secondary electrons. The impact of magnetic fields on the transport of secondary electrons has been well studied in the physics literature, starting more than 50 years ago (Bostick 1950). Recent studies have employed Monte Carlo simulation (Bielajew 1993; Earl and Ma 2002; Jette 2000a, b, 2001; Litzenberg et al. 2001; Naqvi et al. 2001; Nath and Schulz 1978; Shih 1975; Wadi-Ramahi et al. 2001, 2003; Weinhous et al. 1985) and analytic analysis (Jette 2000a) in an attempt to use a localized magnetic field to help focus or trap primary or secondary electrons to increase the local dose deposition in the patient. For high-field MRI with magnetic fields between 1.5 and 3.0 T, it is known that the initial radius of gyration is small with respect to the mean free path of large-angle scattering interactions for the secondary electrons (bremsstrahlung, elastic scatter, and hard collisions), and this condition results in the trapping or focusing of the secondary electrons. This same effect can reduce the quality of a megavoltage photon radiotherapy beam. Indeed, it has already been shown in the literature that a 1.5-T MRI magnetic field perpendicular to a 6-MV beam can both degrade the penumbra (Raaymakers et al. 2004) and cause high exit dose hot spots near any tissue interface to low-density materials (Raaijmakers et al. 2005). Some of these results are shown in Figure 12.5.

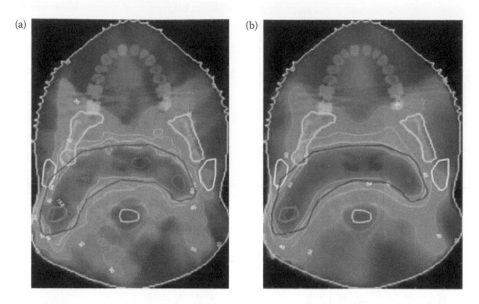

FIGURE 12.3 **(See color insert.)** A Head and Neck treatment plan using 7 equally spaced fields for (a) 6 MV and (b) cobalt.

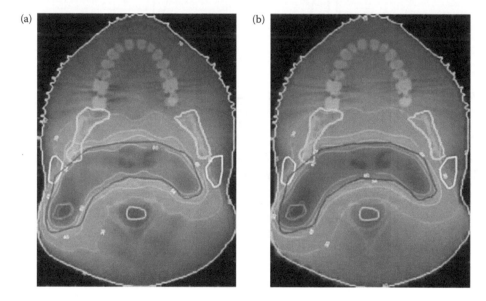

FIGURE 12.4 **(See color insert.)** A Head and Neck treatment plan using 71 equally spaced fields for (a) 6 MV and (b) cobalt.

In the ViewRay™ device, the magnetic field is orthogonal to the radiation beams. To avoid significant dose deposition distortions and to prevent MRI artifacts that could compromise the spatial integrity of the imaging data, ViewRay proposed using a low-field open MRI design that allowed the magnetic field to be directed along the superior–inferior direction of the patient. The Lorentz force, proportional to the magnetic field magnitude $|\bar{B}|$, causes the radius of gyration to be inversely proportional to the magnetic field (a 1 MeV electron has a radius of gyration of 1.3 cm in a 0.3 T field and 0.34 cm in a 1.5 T field).

A comparison of the impact of low- and high-field MRI on photon dose deposition was conducted using Monte Carlo modeling (Han et al. 1987; Mora et al. 1999; Sichani and Sohrabpour

2004) of a beamlet from a cobalt γ-ray source in a slab phantom geometry using the well-validated (for general electron transport, though not explicitly in magnetic fields) (Schaart et al. 2002) Integrated Tiger Series (ITS) Monte Carlo package and its ACCEPTM subroutine for transport in magnetic fields (Halbleib et al. 1994). For the simulations, 0.1 MeV electron and 0.01 MeV photon transport energy cutoffs were imposed, the standard condensed history energy grid (ETRAN approach), energy straggling was sampled from Landau distributions, the mass-collisional stopping powers were based on Bethe theory, and incoherent scattering including binding effect were included. Three pairs of simulations were run in which each pair included the run with and without a 0.3 T uniform magnetic

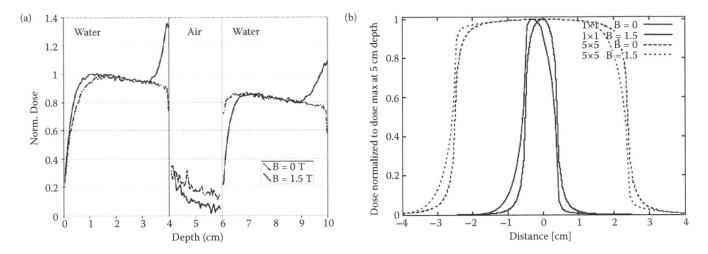

FIGURE 12.5 Significant distortions on 6 MV linac beams from a 1.5 T perpendicular magnetic field: (a) unacceptable distortions of the penumbra. (Adapted from Webb, S., *Intensity Modulated Radiation Therapy*, Institute of Physics Publishing, Philadelphia, 2001a.) And (b) unacceptable hot spots at exiting air tissue interfaces. (Adapted from Webb, S., *Phys. Med.*, 17, 207–215 2001b.)

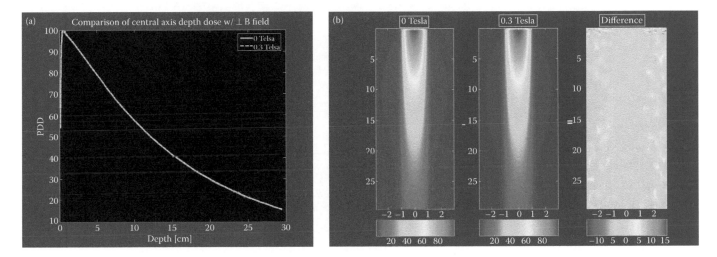

FIGURE 12.6 **(See color insert.)** Comparison of (a) central axis depth dose, and (b) 2D distribution of a 2 cm circular beamlet striking a water phantom with and without a 0.3 T magnetic field applied perpendicular to the beam direction.

field perpendicular to the beam direction. The pencil beams consisted of a 2-cm circular cobalt γ-ray beamlet incident on a 30 × 30 × 30 cm³ water phantom, the phantom with a 10-cm lung density (0.2 g/cc) water slab at 5-cm depth; and a phantom with a 10-cm air density (0.002 g/cc) water slab at 5-cm depth. Simulations were run with between 30 and 100 million histories to obtain <1% σ for all points.

The results are displayed in Figures 12.6 through 12.8. Figure 12.6 clearly demonstrates that a 0.3-T perpendicular uniform magnetic field will not significantly perturb the dose distribution in soft tissue or bone. As seen in Figure 12.7, adding a 10-cm lung density (0.2 g/cc) water slab to the phantom causes a small but noticeable perturbation at the interfaces of the high- and low-density regions. In Figure 12.8, the perturbations are larger and exist mainly in the low-density and interface regions. This demonstrates that air cavities will hold the greatest challenge for

accurate dosimetry (although dose in the air itself has no clinical significance), followed by lung and sinus tissue for low-field MRI radiation therapy. Finally, no hot spots were observed at the air–water interface of the phantom, so the 0.3-T field of the Renaissance™ will not significantly degrade dose distribution quality. Other than at interfaces with lower density media, there should be no perturbations in soft tissue or bone (where the electron mean-free path shortens even more than soft tissue).

12.4.2 The ViewRay™ System

ViewRay expects to launch the world's first real-time MRI-based IGRT system. The system will utilize cutting-edge MRI technology for real-time imaging and a sophisticated MLC collimator and beam delivery system. This turnkey solution will provide comprehensive treatment planning services and will process

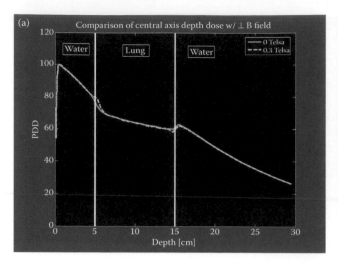
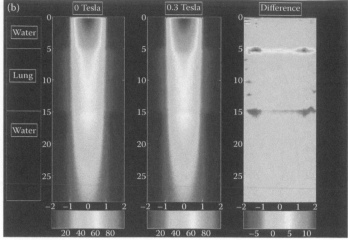

FIGURE 12.7 **(See color insert.)** Comparison of (a) central axis depth dose, and (b) 2D distribution of a 2 cm circular beamlet striking a water/lung/water (5/10/15 cm, respectively) phantom with and without a 0.3 T magnetic field applied perpendicular to the beam direction.

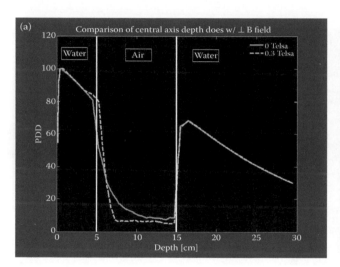
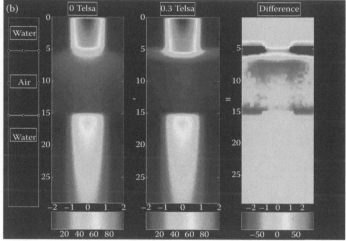

FIGURE 12.8 **(See color insert.)** Comparison of (a) central axis depth dose, and (b) 2D distribution of a 2-cm circular beamlet striking a water/air/water (5/10/15 cm, respectively) phantom with and without a 0.3-T magnetic field applied perpendicular to the beam direction.

image and dose data at high speeds to enable anatomy-based gating and on-table adaptive radiotherapy. The system is expected to incorporate the best features of today's newest IMRT and IGRT systems, with radiation delivery equal to or better than the best competitive system, and true real-time imaging of the tumor during actual beam-on treatment.

The ViewRay™ device consists of the combination of an open split-solenoid MRI scanner equipped for parallel imaging and a ^{60}Co γ-ray IMRT unit. The open configuration of the MRI unit provides a vertical gap located at the field of view. This allows axial beam access to the patient for a gantry mounted -ray IMRT unit to deliver cone-beam MLC-based IMRT. The device is configured so that the isocenter of the IMRT unit and the center of the MRI field of view coincide. Thus, MRI imaging simultaneous with radiation delivery will provide the imaging data necessary to align the patient tissues to their intended positions prior to and

during radiation therapy delivery. The MRI scanner is designed to be a low-field unit (~0.3 T) to allow for imaging with spatial integrity by limiting magnetic susceptibility artifacts due to the patient and to prevent significant perturbations of the dose distribution. An early design of the open MRI and ^{60}Co γ-ray IMRT units are shown in Figure 12.9. The open MRI vertical gap allows for axial beam access to the patient by a ring gantry-mounted ^{60}Co γ-ray IMRT unit that can rotate in a coplanar fashion about the axial gap to deliver static cone-beam MLC-based IMRT.

The current design shown in Figure 12.10 calls for three ^{60}Co γ-ray IMRT to be mounted on the gantry at 120° spacing to increase the dose rate and subsequently decrease the amount of imaging data required by the system. The open MRI is to be equipped with a thin cylindrical gradient system and a parallel phased array RF transceiver coil to allow fast volumetric MRI before, simultaneous to, and after treatment. To enable the true

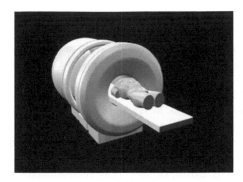
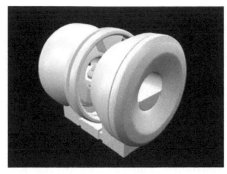
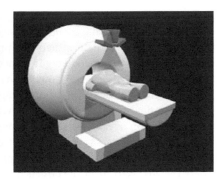

FIGURE 12.9 Three views of the Renaissance™ prototype. (a) A supine patient can be seen entering the device. (b) The device seen from an oblique angle with the ^{60}Co γ-ray IMRT unit in position for an anterior–posterior beam. (c) The device with one of the magnet solenoids, the gantry, and source housing cut away.

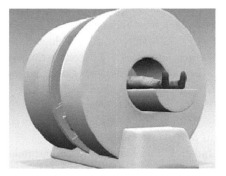

FIGURE 12.10 Three views of the Renaissance™ prototype. (a) A supine patient can be seen entering the device. (b) The device seen from an oblique angle with the ^{60}Co γ-ray IMRT unit in position for an anterior–posterior beam while the couch and one magnet are cut away. (c) The device with one of the magnet solenoids, the gantry, couch, and source housing cut away.

benefit of image-guided radiation therapy the system is designed to rapidly perform four steps:

1. Perform day-to-day and real-time anatomy tracking; that is, know where the patient's radiotherapy targets and anatomy are at all times during every moment of beam-on time through repeated multiplanar and volumetric imaging and deformable image registration.
2. Gate the radiation beam on and off in real time based on actual soft tissue visualization.
3. Compute the dose to the moving patient by using the deformation registration model and the principle of superposition.
4. Reoptimize the fluences delivered to the patient to account for actual changes in the patient geometry and anatomy.

With the ViewRay system, gated control of radiation therapy is achieved through real-time tracking of soft tissue via fast planar MR imaging. Three orthogonal soft-tissue imaging planes can be captured simultaneously and continuously at four frames per second during treatment. Deformable image registration is performed pair wise from the previous to the current frame, tracking voxels and autocontouring the target. The TPDS software allows clinicians to set both spatial and temporal thresholds for pausing treatment delivery. The spatial threshold is applied as a margin to the PTV, while the temporal threshold waits a specified time before terminating irradiation. The tracked target is compared to a user-defined planning target volume contour in the imaged planes. If the alignment of the target and the treatment beams is in violation of both spatial and temporal thresholds, treatment is automatically paused. A wait time is also specified by the user. If the target should move back into the target zone within the wait time, treatment automatically resumes. If the target does not return within the specified wait time, the treatment is paused and the operator must restart. The device is designed to provide a subsecond latency for motion detection and source exposure termination. The device is designed in a fail-safe manner to prevent turning the beam on and off more rapidly than is achievable. Additionally, the system is designed to detect loss of imaging or sudden changes in image acquisition and turn off the beam.

The main goal of the system is to eliminate the limitations imposed on radiation therapy by intra and interfraction organ motion throughout the many weeks of radiation therapy by closing the loop between steps 1, 2, 3, and 4 mentioned above. This will allow the user to compute and know the dose that critical structures and radiotherapy targets receive from each fraction of radiotherapy. Due to the inherent compatibility of MRI and Cobalt radiotherapy, rapid and repeated volumetric imaging simultaneous to radiation delivery is possible. As long as the temporally resolved images maintain spatial integrity, the dose to the patient may be determined by a precise knowledge of the history of the patient geometry in concert with the motion of the MLC leaves. Precisely knowing the dose distribution the patient receives in each fraction of therapy will allow physicians to assess and correct for any gross delivery errors observed due to organ motion over a course of therapy.

If the system works as designed, treatment planning will move from the current 20–35 fraction plan to actual daily plans where the goal is to cover the targets daily and minimize dose to critical structures. Daily optimized plans combined with anatomic gating will allow for margin reductions that could reduce complication rates. Cumulative doses in rigid critical structures can be directly assessed for late effects as well. Deformable critical structures will represent a challenge that will depend on the organ (e.g., adding dose from different deliveries to a spinal cord will be highly accurate due to its lack of mobility while the same procedure for small bowel may be highly inaccurate), but the user will know that the delivery will be as conformal as technically feasible on a daily basis. The faster the loop between the four steps can be closed, the better one will be able to account for organ motion.

References

W. Beavis, P. Gibbs, R. A. Dealey and V. J. Whitton, Radiotherapy treatment planning of brain tumours using MRI alone, *Br J Radiol* **71**, 544–548, 1998.

J. Bernier, E. J. Hall and A. Giaccia, Radiation oncology: a century of achievements, *Nat Rev Cancer* **4**, 737–747, 2004.

A. F. Bielajew, The effect of strong longitudinal magnetic fields on dose deposition from electron and photon beams, *Med Phys* **20**, 1171–1179, 1993.

W. H. Bostick, Possible Techniques in Direct-Electron-Beam Tumor Therapy, *Phys. Rev.* **77**, 564, 1950.

K. S. Chao, J. O. Deasy, J. Markman, J. Haynie, C. A. Perez, J. A. Purdy and D. A. Low, A prospective study of salivary function sparing in patients with head-and-neck cancers receiving intensity-modulated or three-dimensional radiation therapy: initial results, *Int J Radiat Oncol Biol Phys* **49**, 907–916, 2001a.

K. S. Chao, N. Majhail, C. J. Huang, J. R. Simpson, C. A. Perez, B. Haughey and G. Spector, Intensity-modulated radiation therapy reduces late salivary toxicity without compromising tumor control in patients with oropharyngeal carcinoma: a comparison with conventional techniques, *Radiother Oncol* **61**, 275–280, 2001b.

L. Chen, R. A. Price, Jr., L. Wang, J. Li, L. Qin, S. McNeeley, C. M. Ma, G. M. Freedman and A. Pollack, MRI-based treatment planning for radiotherapy: dosimetric verification for prostate IMRT, *Int J Radiat Oncol Biol Phys* **60**, 636–647, 2004.

M. A. Earl and L. Ma, Depth dose enhancement of electron beams subject to external uniform longitudinal magnetic fields: a Monte Carlo study, *Med Phys* **29**, 484–491, 2002.

A. Eisbruch, R. K. Ten Haken, H. M. Kim, L. H. Marsh and J. A. Ship, Dose, volume, and function relationships in parotid salivary glands following conformal and intensity-modulated irradiation of head and neck cancer, *Int J Radiat Oncol Biol Phys* **45**, 577–587, 1999.

L. J. Forrest, T. R. Mackie, K. Ruchala, M. Turek, J. Kapatoes, H. Jaradat, S. Hui, J. Balog, D. M. Vail and M. P. Mehta, The utility of megavoltage computed tomography images from a helical tomotherapy system for setup verification purposes, *Int J Radiat Oncol Biol Phys* **60**, 1639–1644, 2004.

C. Fox, H.E. Romeijn, B. Lynch, C. Men, D.M. Aleman, and J.F. Dempsey, Comparative analysis of 60Co intensity-modulated radiation therapy, *Phys Med Biol* **53**, 3175–3188, 2008.

M. Goitein, Organ and tumor motion: an overview, *Semin Radiat Oncol* **14**, 2–9, 2004.

E. M. Haacke, *Magnetic Resonance Imaging: Physical Principles and Sequence Design.* Wiley, New York, 1999.

J. A. Halbleib, R. P. Kensek, T. A. Mehlhorn, G. Z. Valdez, S. M. Seltzer and M. J. Berger, *Integrated TIGER Series Coupled Electron/Photon Monte Carlo Transport Code System, Version 3.0, RISCC Code Package CCC-467*, edited by D. b. t. R. S. I. Center, 1994.

E. J. Hall and A. J. Giaccia, *Radiobiology for the Radiologist*, 6th ed. Lippincott Williams & Wilkins, Philadelphia, 2006.

K. Han, D. Ballon, C. Chui and R. Mohan, Monte Carlo simulation of a cobalt-60 beam, *Med Phys* **14**, 414–419, 1987.

G. N. Hounsfield, Computerized transverse axial scanning (tomography). 1. Description of system, *Br J Radiol* **46**, 1016–1022, 1973.

G. N. Hounsfield, Picture quality of computed tomography, *AJR Am J Roentgenol* **127**, 3–9, 1976.

G. N. Hounsfield, The E.M.I. scanner, *Proc R Soc Lond B Biol Sci* **195**, 281–289, 1977.

ICRU-50 Report 50, Prescribing, Recording and Reporting Photon Beam Therapy, *Report 50 of the International Commission on Radiation Units and Measurements*, Bethesda, Maryland, 1993.

ICRU-62, Prescribing, Recording and Reporting Photon Beam Therapy (Supplement to ICRU Report 50), *Report 50 of the International Commission on Radiation Units and Measurements*, Bethesda, Maryland, 1999.

IMRTCWG (Intensity-Modulated Radiation Therapy Collaborative Working Group), Intensity-modulated radiotherapy: current status and issues of interest, *Int J Radiat Oncol Biol Phys* **51**, 880–914, 2001.

D. Jette, Magnetic fields with photon beams: Monte Carlo calculations for a model magnetic field, *Med Phys* **27**, 2726–2738, 2000a.

D. Jette, Magnetic fields with photon beams: dose calculation using electron multiple-scattering theory, *Med Phys* **27**, 1705–1716, 2000b.

D. Jette, Magnetic fields with photon beams: use of circular current loops, *Med Phys* **28**, 2129–2138, 2001.

P. Kallman, B. Lind, A. Eklof and A. Brahme, Shaping of arbitrary dose distributions by dynamic multileaf collimation, *Phys Med Biol* **33**, 1291–1300, 1988.

V. S. Khoo, A. R. Padhani, S. F. Tanner, D. J. Finnigan, M. O. Leach and D. P. Dearnaley, Deformable image registration for the use of magnetic resonance spectroscopy in prostate treatment planning, *Int J Radiat Oncol Biol Phys* **58**, 1577–1583, 2004.

R. C. Krempien, K. Schubert, D. Zierhut, M. C. Steckner, M. Treiber, W. Harms, U. Mende, D. Latz, M. Wannenmacher and F. Wenz, Open low-field magnetic resonance imaging in radiation therapy treatment planning, *Int J Radiat Oncol Biol Phys* **53**, 1350–1360, 2002.

K. M. Langen and D. T. Jones, Organ motion and its management, *Int J Radiat Oncol Biol Phys* **50**, 265–278, 2001.

J. S. Laughlin, R. Mohan and G. J. Kutcher, Choice of optimum megavoltage for accelerators for photon beam treatment, *Int J Radiat Oncol Biol Phys* **12**, 1551–1557, 1986.

P. C. Lauterbur, Image formation by induced local interactions: examples of employing nuclear magnetic resonance, *Nature* **242**, 190–191, 1973.

P. C. Lauterbur, Magnetic resonance zeugmatography, *Pure Appl Chem* **40**, 149–157, 1974.

P. C. Lauterbur, C. M. Lai, J. A. Frank and C. S. Dulcey, In vivo zeugmatographic imaging of tumors, abstract, in *Fourth International Conference on Medical Physics,* Ottawa, Canada, 1976, pp. 25–30.

P. C. Lauterbur, M. H. Mendonca Dias and A. M. Rubin, Augmentation of tissue proton spin-lattice relaxation rates by in vivo addition of paramagnetic ions, in *Frontiers of Biological Energetics,* edited by P. O. Dutton, J. Leigh and A. Scarpa, Academic Press, New York, 1978, pp. 752–759.

Y. K. Lee, M. Bollet, G. Charles-Edwards, M. A. Flower, M. O. Leach, H. McNair, E. Moore, C. Rowbottom and S. Webb, Radiotherapy treatment planning of prostate cancer using magnetic resonance imaging alone, *Radiother Oncol* **66**, 203–216, 2003.

D. Letourneau, J. W. Wong, M. Oldham, M. Gulam, L. Watt, D. A. Jaffray, J. H. Siewerdsen and A. A. Martinez, Cone-beam-CT guided radiation therapy: technical implementation, *Radiother Oncol* **75**, 279–286, 2005.

P. Litt, *Isotopes and innovation: MDS Nordion's first fifty years, 1946–1996.* Published for MDS Nordion by McGill-Queen's University Press, Montreal; Ithaca, 2000.

D. W. Litzenberg, B. A. Fraass, D. L. McShan, T. W. O'Donnell, D. A. Roberts, F. D. Becchetti, A. F. Bielajew and J. M. Moran, An apparatus for applying strong longitudinal magnetic fields to clinical photon and electron beams, *Phys Med Biol* **46**, N105–115, 2001.

D. Mah, M. Steckner, A. Hanlon, G. Freedman, B. Milestone, R. Mitra, H. Shukla, B. Movsas, E. Horwitz, P. P. Vaisanen and G. E. Hanks, MRI simulation: effect of gradient distortions on three-dimensional prostate cancer plans, *Int J Radiat Oncol Biol Phys* **53**, 757–765, 2002a.

D. Mah, M. Steckner, E. Palacio, R. Mitra, T. Richardson and G. E. Hanks, Characteristics and quality assurance of a dedicated open 0.23 T MRI for radiation therapy simulation, *Med Phys* **29**, 2541–2547, 2002b.

P. Mansfield, Multi-planar image formation using NMR spin echoes, *J. Phys. C: Solid State Phys* **10**, L55–58, 1977.

P. Mansfield and A. A. Maudsley, Planar spin imaging by NMR, *J Phys. C: Solid State Phys* **9**, L409–411, 1976.

T. Mizowaki, N. Araki, Y. Nagata, Y. Negoro, T. Aoki and M. Hiraoka, The use of a permanent magnetic resonance imaging system for radiotherapy treatment planning of bone metastases, *Int J Radiat Oncol Biol Phys* **49**, 605–611, 2001.

T. Mizowaki, Y. Nagata, K. Okajima, M. Kokubo, Y. Negoro, N. Araki and M. Hiraoka, Reproducibility of geometric distortion in magnetic resonance imaging based on phantom studies, *Radiother Oncol* **57**, 237–242, 2000.

T. Mizowaki, Y. Nagata, K. Okajima, R. Murata, M. Yamamoto, M. Kokubo, M. Hiraoka and M. Abe, Development of an MR simulator: Experimental verification of geometric distortion and clinical application, *Radiology* **199**, 855–860, 1996.

G. M. Mora, A. Maio and D. W. Rogers, Monte Carlo simulation of a typical 60Co therapy source, *Med Phys* **26**, 2494–2502, 1999.

S. A. Naqvi, X. A. Li, S. W. Ramahi, J. C. Chu and S. J. Ye, Reducing loss in lateral charged-particle equilibrium due to air cavities present in x-ray irradiated media by using longitudinal magnetic fields, *Med Phys* **28**, 603–611, 2001.

R. Nath and R. J. Schulz, Modification of electron-beam dose distributions by transverse magnetic fields, *Med Phys* **5**, 226–230, 1978.

B. Petersch, J. Bogner, A. Fransson, T. Lorang and R. Potter, Effects of geometric distortion in 0.2T MRI on radiotherapy treatment planning of prostate cancer, *Radiother Oncol* **71**, 55–64, 2004.

J. Pouliot, A. Bani-Hashemi, J. Chen, M. Svatos, F. Ghelmansarai, M. Mitschke, M. Aubin, P. Xia, O. Morin, K. Bucci, M. Roach, 3rd, P. Hernandez, Z. Zheng, D. Hristov and L. Verhey, Low-dose megavoltage cone-beam CT for radiation therapy, *Int J Radiat Oncol Biol Phys* **61**, 552–560, 2005.

F. J. Prott, U. Haverkamp, H. Eich, A. Resch, O. Micke, A. R. Fischedick, N. Willich and R. Potter, Effect of distortions and asymmetry in MR images on radiotherapeutic treatment planning, *Int J Cancer* **90**, 46–50, 2000.

B. W. Raaymakers, A. J. Raaijmakers, A. N. Kotte, D. Jette and J. J. Lagendijk, Integrating a MRI scanner with a 6 MV radiotherapy accelerator: dose deposition in a transverse magnetic field, *Phys Med Biol* **49**, 4109–4118, 2004.

A. J. Raaijmakers, B. W. Raaymakers and J. J. Lagendijk, Integrating a MRI scanner with a 6 MV radiotherapy accelerator: dose increase at tissue-air interfaces in a lateral magnetic field due to returning electrons, *Phys Med Biol* **50**, 1363–1376, 2005.

H. E. Romeijn, R. K. Ahuja and J. F. Dempsey, A new linear programming approach to radiation therapy treatment planning problems, *Oper Res* **54**, 201–216, 2006.

H. E. Romeijn, R. K. Ahuja, J. F. Dempsey, A. Kumar and J. G. Li, A novel linear programming approach to fluence map optimization for intensity modulated radiation therapy treatment planning, *Phys Med Biol* **48**, 3521–3542, 2003.

H. E. Romeijn, J. F. Dempsey and J. G. Li, A unifying framework for multi-criteria fluence map optimization models, *Phys Med Biol* **49**, 1991–2013, 2004.

D. R. Schaart, J. T. Jansen, J. Zoetelief and P. F. de Leege, A comparison of MCNP4C electron transport with ITS 3.0 and experiment at incident energies between 100 keV and 20 MeV: influence of voxel size, substeps and energy indexing algorithm, *Phys Med Biol* **47**, 1459–1484, 2002.

D. M. Shepard, M. C. Ferris, G. H. Olivera and T. R. Mackie, Optimizing the delivery of radiation therapy to cancer patients, Siam Review **41**, 721–744, 1999.

C. C. Shih, High energy electron radiotherapy in a magnetic field, *Med Phys* **2**, 9–13, 1975.

S. Shimizu, H. Shirato, H. Aoyama, S. Hashimoto, T. Nishioka, A. Yamazaki, K. Kagei and K. Miyasaka, High-speed magnetic resonance imaging for four-dimensional treatment planning of conformal radiotherapy of moving body tumors, *Int J Radiat Oncol Biol Phys* **48**, 471–474, 2000.

B. T. Sichani and M. Sohrabpour, Monte Carlo dose calculations for radiotherapy machines: Theratron 780-C teletherapy case study, *Phys Med Biol* **49**, 807–818, 2004.

R. J. Steenbakkers, K. E. Deurloo, P. J. Nowak, J. V. Lebesque, M. van Herk and C. R. Rasch, Reduction of dose delivered to the rectum and bulb of the penis using MRI delineation for radiotherapy of the prostate, *Int J Radiat Oncol Biol Phys* **57**, 1269–1279, 2003.

H. D. Suit, Choice of optimum megavoltage for accelerators for photon beam treatment, *Int J Radiat Oncol Biol Phys* **12**, 1711–1712, 1986.

B. P. Sutton, D. C. Noll and J. A. Fessler, Fast, iterative image reconstruction for MRI in the presence of field inhomogeneities, *IEEE Trans Med Imaging* **22**, 178–188, 2003.

S. F. Tanner, D. J. Finnigan, V. S. Khoo, P. Mayles, D. P. Dearnaley and M. O. Leach, Radiotherapy planning of the pelvis using distortion corrected MR images: the removal of system distortions, *Phys Med Biol* **45**, 2117–2132, 2000.

N. J. Taylor, H. Baddeley, K. A. Goodchild, M. E. Powell, M. Thoumine, L. A. Culver, J. J. Stirling, M. I. Saunders, P. J. Hoskin, H. Phillips, A. R. Padhani and J. R. Griffiths, BOLD MRI of human tumor oxygenation during carbogen breathing, *J Magn Reson Imaging* **14**, 156–163, 2001.

S. D. Thomas, M. Mackenzie, D. W. Rogers and B. G. Fallone, A Monte Carlo derived TG-51 equivalent calibration for helical tomotherapy, *Med Phys* **32**, 1346–1353, 2005.

S. J. Wadi-Ramahi, D. Bernard and J. C. Chu, Effect of ethmoid sinus cavity on dose distribution at interface and how to correct for it: magnetic field with photon beams, *Med Phys* **30**, 1556–1565, 2003.

S. J. Wadi-Ramahi, S. A. Naqvi and J. C. Chu, Evaluating the effectiveness of a longitudinal magnetic field in reducing underdosing of the regions around upper respiratory cavities irradiated with photon beams--a Monte Carlo study, *Med Phys* **28**, 1711–1717, 2001.

A. P. Warrington and E. J. Adams, presented at the Proceedings of the 6th Biennial Physics Meeting of the European Society for Therapeutic Radiation Oncology, Sevilla, Spain, 2001 (unpublished).

S. Webb, *Intensity Modulated Radiation Therapy*. Institute of Physics Publishing, Philadelphia, 2001a.

S. Webb, The future of photon external-beam radiotherapy the dream and the reality, *Physica Medica* **17**, 207–215, 2001b.

M. S. Weinhous, R. Nath and R. J. Schulz, Enhancement of electron beam dose distributions by longitudinal magnetic fields: Monte Carlo simulations and magnet system optimization, *Med Phys* **12**, 598–603, 1985.

X. Wu, S. J. Dibiase, R. Gullapalli and C. X. Yu, Deformable image registration for the use of magnetic resonance spectroscopy in prostate treatment planning, *Int J Radiat Oncol Biol Phys* **58**, 1577–1583, 2004.

M. J. Zelefsky, Z. Fuks, L. Happersett, H. J. Lee, C. C. Ling, C. M. Burman, M. Hunt, T. Wolfe, E. S. Venkatraman, A. Jackson, M. Skwarchuk and S. A. Leibel, Clinical experience with intensity modulated radiation therapy (IMRT) in prostate cancer, *Radiother Oncol* **55**, 241–249, 2000.

M. J. Zelefsky, Z. Fuks, M. Hunt, H. J. Lee, D. Lombardi, C. C. Ling, V. E. Reuter, E. S. Venkatraman and S. A. Leibel, High dose radiation delivered by intensity modulated conformal radiotherapy improves the outcome of localized prostate cancer, *J Urol* **166**, 876–881, 2001.

Fault Detection in Image-Based Tracking

13.1	The Hokkaido RTRT System	129
13.2	Tracking, Prediction, and Online Monitoring	130
13.3	Human Factors	130
13.4	Dependable Systems	131
13.5	Theory of Reliable Systems—Hardware	131
	Failure Detection • Failure Recovery	
13.6	Theory of Reliable Systems—Software	133
	Fault Detection • Failure Recovery	
13.7	System Verification and Validation	134
	References	134

Gregory C. Sharp

Rui Li

Nagarajan Kandasamy

The purpose of a fault-detection system is to identify abnormal system behavior. In the context of an image-based tracking system, we focus on the subsystems responsible for target localization because loss of track is the most common form of fault. In principle, loss of track could be detected entirely by a dedicated operator who monitors the tracking software in real time. However, it is desirable to provide monitoring software which assists the operator in this task. Online monitoring software reacts more quickly than human operators and improves the repeatability and consistency of the task.

13.1 The Hokkaido RTRT System

In this section, we describe the design of the online monitoring system used in the Hokkaido University real-time tumor-tracking radiation therapy (RTRT) system.

The RTRT system consists of a medical linear accelerator and a room-mounted stereoscopic x-ray fluoroscopy system. Metallic gold markers are implanted in or near the tumor site, and the positions of these markers are visible in the fluoroscopic video. The relative position of the markers relative to the treatment site is computed from CT and stored in the delivery system computer. At the time of treatment, patients are positioned on the treatment couch so that the tumor is targeted by the treatment beam.

Once the patient is positioned, the fluoroscopic imaging system is engaged, and pattern recognition software is used to track the position of the gold marker over time. Because the marker is viewed stereoscopically from two separate viewing directions, the three-dimensional (3D) coordinates of the marker are known. When the marker is located within the preplanned location for treatment, the accelerator delivers radiation to the target. When the marker is outside of the preplanned location, radiation is not delivered. Thus, the system performs respiratory gated treatment (cf. Chapter 3) based on the internal anatomy, as seen by the fluoroscopic tracking system.

The tracking software uses normalized cross-correlation (NCC) to detect the position of the marker. As mentioned previously, this metric is robust to changes in the image brightness but can be misled by variations in the image background. Each imaging subsystem reports a single two-dimensional (2D) marker location, which is the maximum of the NCC score within an operator-specified region of interest. In addition to reporting the 2D marker location, each system reports the pattern recognition score (PRS), which is a measure of confidence in the marker location.

After the marker has been located in both imagers, geometry calibration data are used to compute the 3D location using a ray intersection method. For each imager, an imaginary ray is drawn from the imaging source to the 2D marker location on the imager. If the 2D marker detection is perfect, the two rays will intersect exactly at the 3D location of the marker. However, in general, the rays do not exactly match. Therefore, in addition to computing the 3D position of the marker, the triangulation software reports distance between rays (DBR) at their closest point. When the DBR is large, it is unlikely that the marker location is accurate.

In total, there are five adjustable thresholds which can be set to control fault detection in the RTRT system, which are described in Table 13.1 and illustrated in Figure 13.1. The system has one soft threshold, and four hard thresholds. The soft threshold is called the *PRS level*, and the beam is held when the PRS falls below this value. If the PRS remains below the PRS level for more than a fixed number of frames specified by the *PRS Frame Limit*, a system interlock is triggered. System interlocks are also engaged if the PRS falls below the *PRS Interlock Level*, if the target remains outside of the gating window for longer than the *Waiting Time*, or if the DBR falls below the *DBR Level*.

13.2 Tracking, Prediction, and Online Monitoring

It is natural to ask the question as to the distinction between tracking, prediction, and online monitoring. The tracking system at all times tries create the best estimate of the current target position, and the prediction system attempts to estimate the future position. It is customary for these estimates to make use of signal processing methods. For example, a tracking system might use a statistical tracking method such as a Kalman filter or particle filter (cf. Chapter 2), which biases the position estimates toward smooth motions. Prediction systems use similar smoothing filters to predict smooth trajectories. It is of

questionable benefit to design a tracking system that generates smooth trajectories and then to build an online monitoring system that tests the tracker for smooth trajectories.

Instead, it is better to generate accuracy estimates from measurements. In the RTRT system, two such accuracy estimates are available: the distance between rays and the pattern recognition score. Statistical trackers that produce maximum likelihood estimates of position can similarly generate likelihood values for each image frame. The online monitoring system can then use these values, which may be more informative than trajectory smoothness (Figure 13.1).

13.3 Human Factors

The process of fault detection is naturally enhanced by active participation from the machine operator. In the case of image-based tracking systems, one is primarily concerned with safety and reliability. Safety is preserved by ensuring that the treatment is delivered to the correct location and that the imaging dose is kept within reasonable limits. System reliability dictates the additional condition that the radiotherapy treatment be completed promptly.

To make best use of human operators, we must design procedures that use automatic tools for low-level tasks, but give the operators control over the important decisions. An example of this design is shown in Figure 13.2, which shows a hybrid procedure for failure detection. In this scenario, automatic fault detection is used to hold the beam for transient failures or to interlock the device for unrecoverable failures. Simultaneously, the radiation therapist monitors the tracking system and looks for undetected failures. To be effective, a hybrid system must provide high-level access to important tracking information. In particular, there must be a live display of the images, and the estimated target position must be clearly marked. Numeric fault detection cues, such as pattern recognition scores, should also be

TABLE 13.1   Fault Detection Thresholds in the RTRT System

PRS Level	If the PRS falls below this level, the beam is held.
PRS Interlock Level	If the PRS falls below this level, the linac interlock is triggered.
PRS Frame Limit	If the PRS falls below the PRS level for more than this number of image frames, the linac interlock is triggered.
Waiting Time	If the marker position does not enter the gating window for longer than this amount of time, the linac interlock is triggered.
DBR Level	If the DBR exceeds this level, the linac interlock is triggered.

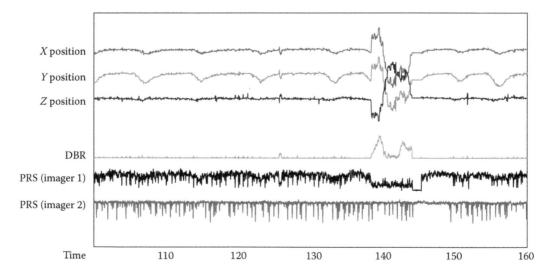

FIGURE 13.1   Online monitoring in the RTRT system. Three statistics are monitored: the pattern recognition score (PRS) for both imagers and the distance between rays (DBR).

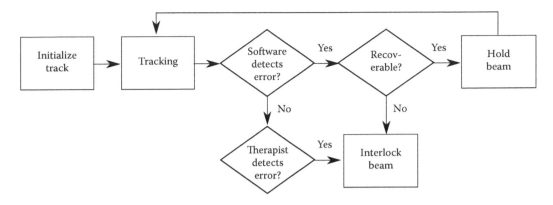

FIGURE 13.2 Process diagram for tracking error detection with a human in the loop.

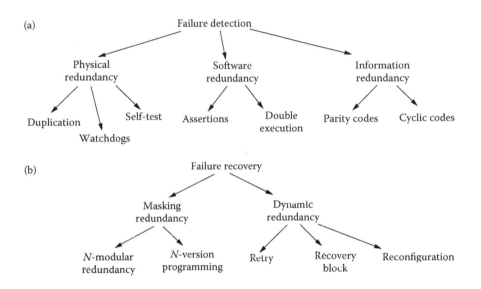

FIGURE 13.3 A classification of some approaches to (a) failure detection, and (b) recovery.

available. Ideally, these values will be graphed relative to historic values or action thresholds so that the operator can verify their correctness.

13.4 Dependable Systems

A dependable computer system provides a function upon which reliance can be justifiably placed [SS88]. Dependability is a generic term applicable to attributes such as safety, reliability, availability, and security, which affect the quality of service provided by the system. In this section, dependability is used in the context of *fault tolerance*, which is defined as the ability of the system to operate correctly even after hardware and software failures occur. A system fails if it delivers a function deviating from the one specified. Most failures are caused by specification, design, and implementation errors; improper operator action; and environmental disturbances such as electromagnetic interference. Such failures are typically classified by their duration. *Permanent failures* remain in existence indefinitely if no corrective action is taken. Although many are caused by residual

design or manufacturing faults, they are also caused by catastrophic events such as an accident. Intermittent faults appear, disappear, and reappear repeatedly. They are difficult to predict, but their effects are highly correlated. Most *intermittent faults* are due to marginal design, testing, or manufacturing, and they manifest themselves under certain environmental or system conditions. *Transient faults* appear and disappear quickly and are not correlated with each other. They are most commonly induced by random environmental disturbances such as electromagnetic interference.

If fault tolerance is required, then some form of redundancy must be applied at additional cost to the system. Figure 13.3a shows a classification of commonly used failure-detection methods.

13.5 Theory of Reliable Systems—Hardware

This section discusses hardware-based approaches to failure detection and recovery.

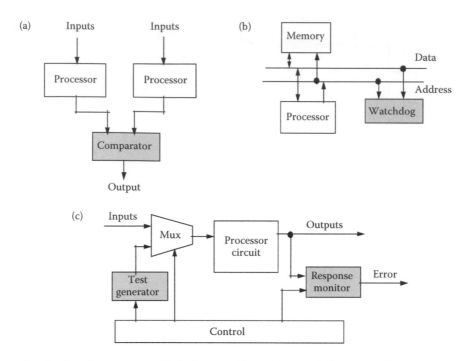

FIGURE 13.4   Hardware-based failure detection using (a) duplication, (b) watchdogs, and (c) built-in self-test.

13.5.1 Failure Detection

Hardware duplication shown in Figure 13.4a is a simple fault-detection technique in which the results of two identical processors are compared. Fault is detected if the results disagree. This procedure detects all single faults, except those of the comparator, at the expense of approximately 100% cost overhead. However, if the processors generate results using identical hardware and software, a design error affecting both processors may cause identical failures not detected by the comparator. Such common-mode failures are avoided by means of design diversity, wherein processors deliver the same service through separate designs and implementations. The overhead due to duplication may be reduced using the simpler *watchdog* mechanism, shown in Figure 13.4b, to continuously monitor processor behavior for errors. For example, a software program is decomposed into blocks or groups of instructions, and a signature is computed for each block. All valid execution sequences of these blocks are stored within the watchdog. As the processor executes instruction blocks, the corresponding signatures are transmitted to the watchdog, which compares the signature stream with the previously stored program-flow information. A mismatch indicates an error. Watchdogs detect control flow and timing errors in the main processor.

The *built-in self-test* approach shown in Figure 13.4c places the testing circuit physically on the same chip as the processor [AAMH98]. The tester applies a sequence of test patterns to processor circuits and evaluates the corresponding responses for failures. Testing may be performed either during system startup or during its operation. Testing features may include parity codes for arithmetic operations, program control-flow checking, watchdog timers, and power-supply monitoring.

13.5.2 Failure Recovery

After a failure is detected, the computing system can sometimes be placed in a safe state with degraded functionality. If, however, a safe system state cannot be identified, the system must continue to operate using some form of redundancy. We now review some commonly used failure recovery methods.

Masking uses hardware redundancy to correct failures before they affect processor outputs. Typically, fault masking provides no fault detection or warning of faulty processors until a minimum number of faults have accumulated. As a result, most masking schemes are extended to provide some fault detection as well [SS88]. Figure 13.5a shows the basic triple modular redundancy (TMR) configuration, in which three processors perform the same computation and the final result is a majority vote on the outputs. TMR can thus mask a single processor failure. Voters that are themselves failure sources in Figure 13.5a can be replicated as shown in Figure 13.5b. Masking can also be applied in straightforward fashion to other components such as sensors and actuators.

Since static redundancy preallocates resources for the worst-case failure scenarios, the cost overhead may be excessive. On the other hand, dynamic redundancy uses a combination of fault detection, reconfiguration, and recovery to tolerate failures in a cost-effective fashion. *Primary/backup* is a well-known method of tolerating failures using dynamic redundancy. In "hot standby" mode, the backup processor operates continuously, and if the primary fails, it is switched on with minimal recovery latency. However, the backup consumes power and ages with continued service. In "cold standby" mode, the backup is inactive, and if the primary fails, the backup enters operation after some startup cost, including that of transferring the operational

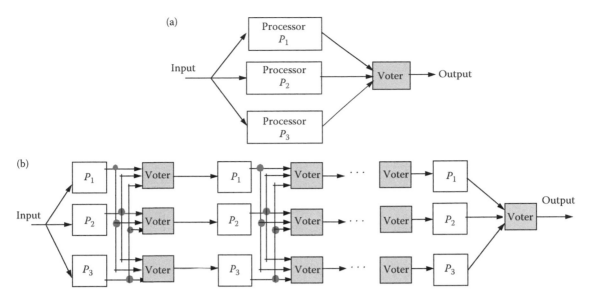

FIGURE 13.5 (a) The basic triple-modular redundancy (TMR) scheme, and (b) a TMR scheme with triplicated voters.

state from the failed processor. In this configuration, the backup does not age or consume power while the primary operates. In both schemes, however, failures affecting the primary must be correctly diagnosed in a timely fashion.

13.6 Theory of Reliable Systems—Software

In many computing systems, software-based methods provide a low-cost alternative to hardware redundancy for both fault detection and recovery.

13.6.1 Fault Detection

Two widely used approaches are assertions and acceptance checks, both of which use application-specific knowledge to detect failures. *Assertions* check if a task satisfies specific constraints at particular points, whereas *acceptance tests* use range, bounds, and sanity checks to verify the result. These tests must be carefully designed and evaluated to achieve high fault coverage while minimizing false alarms. For example, overly sensitive tests can reject the results because of small deviations from optimal performance even though the risk of an accident happening may be small. A more systematic method of detecting transient failures is to simply execute a task twice and compare the results.

13.6.2 Failure Recovery

Masking techniques for hardware failures can be applied to software as well in the form of *N-version programming* (NVP) that uses design diversity to mask task failures [Ed95]. In NVP, two or more versions of the program providing the same result through separate designs and implementations are produced. The final result is then obtained by a majority vote on the

outputs of the N versions. Common-mode failures, the extra cost for redundant software, and the complexity of the voter are the disadvantages of this approach. In Figure 13.5a, the replicated processors receive identical input data—a source of common-mode failures. Diversity in a task's input data is sometimes used as an alternative to the NVP approach. Techniques such as *N-copy programming* require diverse data as inputs for multiple executions of a single version. The various software versions may execute on different processors using hardware redundancy or on the same processor using time redundancy. In low-cost computer systems, the NVP and TMR schemes can also be realized using time redundancy.

We now discuss some software-based recovery approaches using dynamic redundancy. *Task retry* is the simplest and fastest form of recovery and is effective against transient failures. Other commonly used methods include primary/backup and recovery blocks. In Figure 13.6a, the primary version of the task T_i executes first, and if the acceptance test fails, the backup is executed. To maximize the chances of a successful recovery, the backup may use both design and data diversity. Figure 13.6b shows an alternative to straightforward TMR task execution. First, the task T_i is executed twice and the results are compared. If a mismatch occurs, T_i is executed a third time and the final result is obtained by a majority vote on the three results. Since T_i is executed the third time only if a fault is detected during the previous two executions, the overhead associated with a typical TMR scheme is not incurred.

The recovery block approach in Figure 13.6c combines checkpointing with alternate task versions to tolerate software design errors and transient hardware faults [Ed95]. Initially, the primary block is executed, followed by its acceptance test. If the test passes, the results are sent out. Otherwise, alternate blocks are executed in order, and if they all fail, then the recovery block fails. Although each of the alternates tries to satisfy the same acceptance test, they need not all produce the same results.

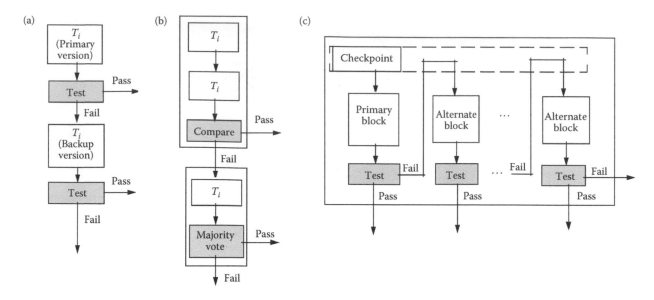

FIGURE 13.6 Software-based approaches to dynamic redundancy: (a) primary/backup, (b) alternate TMR, and (c) recovery blocks.

Therefore, while the primary provides the specified service, the alternates may provide degraded service to the system.

13.7 System Verification and Validation

Having discussed hardware- and software-based fault tolerance techniques, this section discusses the process of system verification and validation [Sto96]. To have confidence in the final system, one must confirm that the following phases of the development work have been performed correctly: requirements, analysis, design, and evaluation. *Verification* is the process of determining whether the output of a development phase actually fulfills the requirements specified by the previous phase. The task of verification is to show that the output of a phase conforms to its input, rather than to show that the output is actually correct. If the input specification to a phase is itself incorrect, then the verification process will not necessarily detect this. So, system verification is often supplemented by *validation*, which is the process of confirming that the specification of a phase (or for a subsystem) is consistent with corresponding user requirements.

Testing is a crucial part of the verification and validation process and can be classified into three major classes: module or unit testing, system integration testing, and system validation testing. *Module* or *unit testing* aims to detect errors in individual software functions or hardware components, whereas *system integration testing* is aimed at establishing the correct interaction between a set of hardware/software components. *System validation testing* aims to show that the complete system satisfies its requirements. The testing process can be static or dynamic in nature. Static testing methods investigate properties of a hardware/software component without actually operating it; for example, via code reviews and inspections and design walkthroughs. In a software context, static techniques include control flow and data flow analysis of programs and formal verification, and these techniques can be used to establish properties of the software that are true under all operating conditions. Dynamic testing requires that the component under test be executed in order to investigate its characteristics. These tests may be conducted during normal operating conditions or under simulated conditions.

References

[AAMH98] H. Al-Asaad, B. T. Murray, and J. P. Hayes. Online BIST for embedded systems. *IEEE Design and Test*, pp. 17–24, October 1998.

[Ed95] M. Lyu (Editor). *Software Fault Tolerance*. John Wiley & Sons, Hoboken, NJ 1995.

[SS88] D. P. Siewiorek and R. S. Swarz. *Reliable Computing Systems: Design and Evaluation*. A. K. Peters, Ltd., 3rd edition, Natick, MA 1988.

[Sto96] N. Storey. *Safety-Critical Computer Systems*. Prentice Hall, Upper Saddle River, NJ 1996.

Index

Note: *Italicized* page references denote figures and tables.

A

Acceptance tests, 133
Accuray, 1
Accurate online support vector regression (AOSVR), 47, 56–57
Active breathing control, 24
Adaptive fractionation therapy, 103
Adaptive radiation therapy (ART), 36, 102
AlignRT® system, 6, 41
Amplitude gating, 25
AquaPlast® mask, 29, 30
Assertions, 133
Audio coaching, 24
Audio–visual feedback-based systems, 24

B

Beam-gated Hokkaido system, 30
Binary multileaf collimator (bMLC), 93–94, 94; see also Real-time motion adaptation
BodyFIX vacuum cushion system, 41, 43
Boyer–Strait–Webb (BSW) construction, 34
BrainLAB AG, 2
BrainLAB ExacTrac 6D system, 2
Breath-hold methods, 23–24
Breathing leaves method, 34
Built-in self-test approach, 132, 132

C

Calypso 4D Localization System, 4
Calypso® System, 41
Candidate patients
    breathing behaviors and respiratory maneuvers
        assessing regularity and stability of respiration, 22–23, 23
        breath-hold, 23–24
        feedback-guided free breathing, 24
        motion restriction, 24
    and treatment sites, 22
CC, see Cross correlation (CC)
Clinical target volume (CTV), 66, 69, 69, 70
Cobalt therapy, 115–116

Cobalt-based treatment system, 4, 5
"Cold standby" mode, 132
Commercial x-ray computed tomography, 33
Commissioning procedure, for gating system, 26
Computed tomography (CT) imaging vs. MRI, for intrafraction organ motion, 119
Conformal radiotherapy (CFRT) techniques, 33, 115–116
Constant linear accelerator dose rate, DMLC control algorithms, 84
Conventional radiosurgery, 29
Correlation, 48
    advanced methods, 48–49
    basic methods, 48
    and prediction, fusion of, 60
    validation experiment, 49
        correlation results, 50–53
        LEDs selection, and using multiple LEDs, 53–54
Couch-based motion correction, in radiation therapy, 44, 44
Couch-based phantom motion compensation, 42
Couch-based target alignment, 39
    couch-based motion correction in radiation therapy, 44, 44
    couch shifts, for patient alignment, 39–41, 41
    dynamic couch-based target motion compensation, 41–43, 43
    quality assurance, 43–44, 44
    technologically advanced treatment couch systems, 40
    transient and low-frequency, 41, 42
    translational shifts, 40
Couch shifts, for patient alignment, 39–41, 41
Cross Cancer Institute, 4
Cross correlation (CC), 16
    zero-mean version of, 17
CTV, see Clinical target volume (CTV)
CyberKnife® image-guided radiosurgery system, 29
    basic configuration, 30
    nonperiodic movement, adaptation to, 29–30
    periodic motion, adaptation to, 30–31, 31

CyberKnife Robotic Radiosurgery System, 2
CyberKnife Synchrony system, 1–2, 2, 9, 31, 47
Cyberknife treatment plans, for pancreatic tumor, 73

D

Data association, 19
DBR, see Distance between rays (DBR)
DBR level, 130, 130
DCE-MRI, see Dynamic contrast enhanced imaging (DCE-MRI)
Deep-inspiration breath-hold method, 24
Deformable template matching, 16
Delivery design tracking, by direct aperture optimization, 35–36, 36
Dempsey, J.F., 119, 120
Dependable computer systems, 131, 131
Detection theory, 13, 15; see also Target detection and tracking, theoretical aspects of
Diffusion weighted imaging (DWI), 107
Digitally reconstructed radiograph (DRR), 31
Direct aperture optimization, in 4D, 35–36, 36
Direct tumor localization methods, 6–8, 7
Distance between rays (DBR), 129, 130
DJDC TomoTherapy delivery, see Dynamic jaw and dynamic couch (DJDC) TomoTherapy delivery
DMLC, see Dynamic multileaf collimator (DMLC)
DMLC IMRT, see Dynamic multileaf collimator intensity-modulated delivery (DMLC IMRT)
Dose–volume histograms (DVHs), 120
Dosimetry, and treatment planning, 108–111, 110, 111
DRR, see Digitally reconstructed radiograph (DRR)
Dual polynomial model, 48
Duty cycle, 23
DVHs, see Dose–volume histograms (DVHs)
DWI, see Diffusion weighted imaging (DWI)

Dynamic contrast enhanced imaging (DCE-MRI), 107
Dynamic couch-based target motion compensation, 41–43, *43*
Dynamic jaw and dynamic couch (DJDC) TomoTherapy delivery, 103
Dynamic multileaf collimator (DMLC), 34, 77
  control algorithms
    with constant linear accelerator dose rate, 84
    evolution of, *79*
    for 1D moving and deforming targets, 79–81, *80*, *81*
    for 3D moving targets, 81–84
    with variable linear accelerator dose rate, 84–85
  motion management via, *see* Motion management via DMLC tracking, treatment planning for
  tracking leaf-sequencing evolution
    algorithm properties, summary of, 79, *79*
    basic governing equations, 78–79
    as tracking solution, 78
Dynamic multileaf collimator intensity-modulated delivery (DMLC IMRT), 77
  complications and issues with, 78
  IMRT delivery to moving targets, 77–78
  as tracking solution, 78
Dynamic redundancy, 132, 133
  software-based approaches to, *134*
Dynamic testing methods, 134

E

Electromagnetic (EM) transponder, 4, 6
  -based localization, 7
Electronic portal imaging device (EPID), 1, 3, 4
Elekta Inc., 3
Elekta MLCi collimator design, 6, 35
Elekta Synergy system, 1, 3, *3*
Empirical template, 16
End-of-exhalation breath hold method, 24
e-support vector regression (e-SVR), 47
ExacTrac 6D system, 1, 2, *2*
Exactrac® Robotics module, *40*
External-beam radiotherapy, 33
External surrogates, 8, 9

F

Failure detection, *131*
  hardware-based approaches, 132, *132*
  software-based approaches, 133
Failure recovery, *131*
  hardware-based approaches, 132–133, *133*
  software-based approaches, 133–134, *134*
Fast magnetic resonance imaging, for real-time tumor localization, 4, *5*
Fault detection, in image-based tracking system, 129

dependable systems, 131, *131*
Hokkaido RTRT system, 129–130, *130*, *130*
  fault detection thresholds in, *130*
  online monitoring in, *130*
human factors, 130–131, *131*
system verification and validation, 134
theory of reliable systems
  hardware-based approaches, 132–133, *132*, *133*
  software-based methods, 133–134, *134*
tracking, prediction, and online monitoring, 130
Fault tolerance, 131
Feedback-guided free breathing, 24
Fiducial markers, 7
Fluoroscopic imaging system, 3, 13, 129
Foreground object, size of, 15
4D, direct aperture optimization in, 35–36, *36*
4D computed tomography (4DCT), 7, 22, 24, 25, 66, 68, 69
4D-internal target volume (4D-ITV), 69, 70
4D phantom, 100, *101*
Frameless cranial radiosurgery, 29
Full synchronization technique, 35

G

Gamma-ray intensity modulated radiation therapy, 119
  treatment plans, 120–121, *120*, *121*, *122*
Gamma ray radiotherapy, 4
Gantry-mounted systems, 1, 3
Gated internal target volume (ITVg), 25
Gated treatment
  delivering, 22, 25–26
  planning, 22, 24–25
  simulating, 24–25
Gating system, 21, 25; *see also* Respiratory gating
  quality assurance and control, 26
    commissioning procedure, 26
    routine quality assurance, 27
Gross target volume (GTV), 66

H

Hammersmith Hospital, MV linear accelerator installation, 115
Hardware-based approaches
  for failure detection, 132, *132*
  for failure recovery, 132–133, *133*
Hardware duplication, 132, *132*
HexaPOD™ evo RT System, *40*
Hokkaido RTRT system, 129–130, *130*, *130*
  fault detection thresholds in, *130*
  online monitoring in, *130*
Hokkaido University Hospital, 1
"Hot standby" mode, 132
Human respiratory motion, prediction of, 56
Human, tracking error detection with, 130–131, *131*
Hybrid optical/x-ray tracking system, 30
Hybrid tumor localization methods, 8–9

I

Image-based tracking system, 13, 15, 129
  dependable systems, 131, *131*
  Hokkaido RTRT system, 129–130
    fault detection thresholds in, *130*
    online monitoring in, *130*
  human factors, 130–131, *131*
  steps in, *18*
  system verification and validation, 134
  theory of reliable systems
    hardware-based approaches, 132–133, *132*, *133*
    software-based methods, 133–134, *134*
  tracking, prediction, and online monitoring, 130
Image guided intensity modulated radiation therapy (IGIMRT), 118
Image-guided radiation therapy (IGRT), 33, 118
  Couch Top System, *40*
  for radiotherapy, CT snapshots, 118–119
Image noise, 14
Image processing, in treatment planning, 66, 68–69
Indiana University, 34
Indirect tumor localization methods, 8
Infrared-based optical positioning system, 2
Integrated radiotherapy imaging system (IRIS), 1, 3, *3*
Intensity-modulated arc therapy, 42
Intensity modulated radiation therapy (IMRT), 33, 42, 77, 93, 116–117
  delivery to moving targets, 77–78
  gamma-ray, 119
    treatment plans, 120–121, *120*, *121*, *122*
  motion-optimized, 84–85
Intermittent faults, 131
Internal surrogates, 8
Internal target volume (ITV), 69
  4D-ITV, 69, 70
  ITVg, 25
Intrafraction breathing motion, 34
Intrafraction (organ) motion, 29, 33
  CT imaging *vs.* MRI for, 119
  radiation therapy, adaptation in, 21
Intratreatment motion adaption, during treatment planning, 72–73, *73*
IRIS, *see* Integrated radiotherapy imaging system (IRIS)

J

Joint probability data association filtering (JPDAF), 19

K

Kalman filter, 130
  correction step, 19
  prediction step, 18
Karush–Kuhn–Tucker (KKT) conditions, 56

Kilovoltage (kV)
    radiographic imaging systems, 1, 14
    fluoroscopy localization, 6

L

LEDs, selection of, 53–54
LibSVM, 48
Likelihood ratio test, 13–14, *14*
Linac multileaf collimator, tracking with, 33;
        see also Robotic LINAC tracking
    adaptive therapy, 36
    direct aperture optimization in 4D,
        tracking delivery design by,
        35–36, *36*
    intrafraction breathing motion, 34
    motion deconvolution attempts, fatal flaw
        of, 35
    one-dimensional motion, tracking with, 34
    two-dimensional motion, tracking with, 35
Longitudinal motion compensation, of MAD
        strategy, 95, *95*, *96*
Lorentz force, 108
Lung tumors, 29, 30, 31
    template-matching procedure, *31*

M

MAD, *see* Motion-adaptive delivery (MAD)
Magnetic field selection, 121–123, *123*, *124*
Magnetic resonance imaging (MRI), 1, 33, 107
    vs. CT imaging, for intrafraction organ
        motion, 119
    and radiotherapy
        magnetic field selection, 121–123,
            123, *124*
        ViewRay™ system, 123–126, *125*
Magnetic resonance linac (MRL),
        107–108, *108*
MAO, *see* Motion-adaptive optimization
        (MAO)
Marker-based direct localization methods,
        6, *7*
Marker implantation, 6, 7
Masking, 132
Massachusetts General Hospital, 3
Matching cost functions, 16
    popular, 16–17
    robust, 17–18
MATLAB®, 48
Maximum of absolute difference, 17
    zero-mean version, 17
Mean image, 16
Mean-squared error tracker, 15
MHT, *see* Multiple hypothesis tracking
        (MHT)
Mid-time synchronization technique, 35
Mitsubishi Electronics, 1
Mitsubishi/Hokkaido real-time tumor-
        tracking (RTRT) system, 1, *2*, 6,
        129–130
    fault detection thresholds in, *130*
    online monitoring in, *130*

MLC, *see* Multileaf collimator (MLC)
Model-based template, 16
Moderate deep-inspiration breath-hold
        method, 24
Module testing, 134
Motion adaptation, in radiation therapy, 65
    image processing, 66, 68–69
    intratreatment motion adaption,
        72–73, *73*
    motion-compensated treatment, planning
        for, 69–72, *69*, *70*, *71*, *72*
    treatment planning, 65–66, *67*, *68*
Motion-adaptive delivery (MAD), 94–95, 97,
        98, 102, 103
    implementation, 97
    longitudinal motion compensation, *95*, *96*
    transversal motion compensation, 95
Motion-adaptive optimization (MAO), 94, 95,
        96–97, 98, 99, 100, *101*, 102, 103
    implementation, 97
    workflow, *96*
Motion-compensated treatment, 42, 43
    treatment planning for, 69–72, *69*, *70*, *71*,
        72
Motion deconvolution attempts, fatal flaw
        of, 35
Motion enhancement, 16, *17*
    images, 16
    and template selection, 15–16, *16*
Motion management via DMLC tracking,
        treatment planning for, 77
    DMLC IMRT, 77–78
    DMLC tracking leaf-sequencing
        evolution, 78–79
    MLC tracking, of moving targets, 88–90
    motion-optimized IMRT, 84–85
    1D moving and deforming targets, DMLC
        control algorithms for, 79–81,
        80, *81*
    3D moving targets, DMLC control
        algorithms for, 81–84
    VMAT
        interdependence of delivery
            parameters of, 90
        motion management in, 86–87
Motion-optimized intensity modulated
        radiation therapy, 84–85
    DMLC control algorithms
        with constant linear accelerator dose
            rate, 84
        with variable linear accelerator dose
            rate, 84–85
Motion restriction, 24
MRI, *see* Magnetic resonance imaging (MRI)
MRI/linac prototype, 4, *5*
MRL, *see* Magnetic resonance linac (MRL)
MULIN family of algorithms, 54–55
Multileaf collimator (MLC), 33, 42, 93
    adaptive therapy, 36
    direct aperture optimization in 4D,
        tracking delivery design by,
        35–36, *36*
    intrafraction breathing motion, 34

motion deconvolution attempts, fatal flaw
        of, 35
    one-dimensional motion, tracking
        with, 34
    tracking of moving targets, basic
        equations for, 88–90
    two-dimensional motion, tracking
        with, 35
Multiobject tracking, 19
    joint probability data association
        filtering, 19
    multiple hypothesis tracking, 19–20
Multiple hypothesis tracking (MHT), 19–20
Megavoltage (MV) radiographic imaging,
        7, 14

N

N-copy programming, 133
Nonperiodic movement, adaptation to, 29–30
Normal inspiration breath-hold method, 24
Normalized cross correlation (NCC), 16, 129
    zero-mean version of, 17
N-version programming (NVP), 133

O

OBI system, *see* Varian onboard imager (OBI)
        system
Object detection, 15, 16
Off-line adaptive radiation therapy, 102
1D moving and deforming targets, DMLC
        control algorithms for, 79–81,
        80, *81*
One-dimensional motion, tracking with, 34
1.5 T MRI accelerator system, *109*
    artistic impression of, *108*
    for real-time image guided radiotherapy,
        107–112
1.5 T Philips Achieva system, 4, 107
Online adaptive radiation therapy, 102
Online monitoring system, 129, 130
    in RTRT system, 130
Optical emission, 1
Organ motions, 117, *118*
Organs at risk (OAR), 66, 70

P

Pancreatic tumor, 29, 30
    Cyberknife treatment plans for, *73*
Parity codes, 132
Particle filter, 19, 130
Patient alignment, couch shifts for, 39–41, *41*
Pattern recognition score (PRS), 129, 130, *130*
Periodic motion, adaptation to, 30–31, *31*
Permanent failures, 131
Peter MacCallum Cancer Center, 4
Phase gating, 25
Planning organ at risk volumes (PRV), 66
Planning target volume (PTV), 33, 66, 117
Popular matching cost functions, 16–17
Positron emission, 1, 4–5, 6, *6*, *7*

Positron emission tomography (PET), 33
Predetermined template, 16
Prediction system, 130
Primary/backup method, 132, 133, *134*
Protura™ 6DOF Robotic Couch, *40*
PRS Frame Limit, 130, *130*
PRS Interlock level, 130, *130*
PRS level, 130, *130*
Pulsatory motion, *see* Respiratory and
        pulsatory motion, prediction of

Q

Quality assurance, in couch-based target
        alignment, 43–44, *44*

R

Radiation therapy, treatment planning in,
        65–66, *67, 68*; *see also* Motion
        adaptation, in radiation therapy
Radio frequency electromagnetic wave, 1
Radiographic imaging, 1–4, *2, 3*
Radiographic kilovoltage x-ray imaging
        system, 2
Radio-opaque marker, 13, 14, *14,* 16
Radiotherapy, 33
    IGRT for, 118–119
    and MRI
        magnetic field selection, 121–123,
            123, 124
        ViewRay™ system, 123–126, *125*
Ray intersection method, 129
Real-time adaptive radiation therapy, 102–103
Real-time 3D ultrasound, 4
Real-time image guided radiotherapy
    linac with 1.5 T MRI for
        clinical impact, 111–112, *111*
        design magnetic resonance linac
            (MRL), 107–108, *108*
        dosimetry and treatment planning,
            108–111, *110, 111*
        status, 108, *109, 110*
Real-time motion adaptation, 93
    binary MLC and TomoTherapy® treatment
        system, 93–94, *94*
    simulations
        clinical data, 98–99, *99, 100*
        synthetic data, 97–98, *97, 98*
    strategies, 94
        motion-adaptive delivery, 94–95,
            95, 96
        motion-adaptive optimization,
            96–97, *96*
        system integration and experiments,
            99–100, *101, 102*
Real-time motion tracking, 30–31
Real-time position management (RPM)
        system, 5, *6*
Real-time synchronized MLC, for targets
        moving in 3D, 82–84
Real-time tumor localization, 1
    methods, 6

direct methods, 6–8, *7*
    hybrid methods, 8–9
    indirect methods, 8
systems
    electromagnetic transponder, 4
    fast MRI, 4, *5*
    positron emission, 4–5, *6*
    radiographic imaging, 1–4, *2, 3*
    respiratory monitoring devices, 5, *6*
    ultrasound, 4
Reconfiguration, 132
Recovery block approach, 133, *134*
Reliable systems, theory of
    hardware-based approaches
        failure detection, 132, *132*
        failure recovery, 132–133, *133*
    software-based methods
        failure recovery, 133–134, *134*
        fault detection, 133
Renaissance™ prototype, views of, *125*
Renaissance system, 4
Respiration, assessing regularity and stability
        of, 22–23, *23*
Respiratory and pulsatory motion, prediction
        of, 54
    MULIN family of algorithms, 54–55
    SVRpred algorithm, 56–57
        AOSVR algorithm, 56–57
        human respiratory motion, prediction
            of, 56
        prediction method, 57
        signal history, selection of, 57
        speed optimization, 57
    validation experiments, 57–59
        batch training and prediction, 59
        evaluation results, 58–59
        parameter selection, 58
        training percentage, influence of, 59
Respiratory gating, 21
    candidate breathing behaviors and
            respiratory maneuvers
        assessing regularity and stability of
            respiration, 22–23, *23*
        breath hold, 23–24
        feedback-guided free breathing, 24
        motion restriction, 24
    candidate patients and treatment sites, 22
    future developments, 27
    gated treatment
        delivering, 25–26
        planning and delivering, 22
        simulating and planning, 24–25
    gating quality assurance and control, 26
        commissioning procedure, for gating
            system, 26
        routine quality assurance, 27
    historical development, 21–22
    limitations, 27
Respiratory monitoring devices, 5, *6*
Respiratory motion, *see* Intrafraction
        breathing motion
Respiratory motion adaptation, *see* Periodic
        motion, adaptation to

Respiratory surrogates, 8
Robotic LINAC tracking, 47, 59–60
    correlation, 48
        advanced correlation methods, 48–49
        basic correlation methods, 48
        validation experiment, 49–54
    prediction and correlation, fusion of, 60
    respiratory and pulsatory motion,
            prediction of, 54
        MULIN family of algorithms, 54–55
        SVRpred algorithm, 56–57
        validation experiments, 57–59
    surrogates to improve prediction
            quality, 61
Robust matching cost functions, 17–18
Romeijn, H.E., 119
Room-mounted x-ray imaging systems, 1, 2–3
Rotational intensity modulated radiation
        therapy, 93
Routine quality assurance, for gating
        system, 27
RPM system, *see* Real-time position
        management (RPM) system
RTRT system, *see* Mitsubishi/Hokkaido
        real-time tumor-tracking (RTRT)
        system

S

SAD, *see* Sum of absolute differences (SAD)
Sensor modeling, 14–15, *15, 15*
Siemens Medical Systems, 5
Signal history, selection of, 57
Signal-to-noise ratio, 15
Single object tracking, 18
    kalman filter, 18–19
    particle filter, 19
Single-photon emission computed
        tomography (SPECT), 33
Software-based methods
    for failure recovery, 133–134, *134*
    for fault detection, 133
Spatial adaptation (tracking), 21
SSD, *see* Sum of squared differences (SSD)
Stanford University, 34
Static beam intensity modulated radiation
        therapy, 93
Static redundancy, 132
Static testing methods, 134
Sum of absolute differences (SAD), 16
    zero-mean version of, 17
Sum of squared differences (SSD), 16
    zero-mean version of, 17
Support vector regression (SVR), 47
Surrogates
    for quality improvement, 61
    tumor localization with, 8
SVR, *see* Support vector regression (SVR)
SVRpred algorithm, 56–57
    AOSVR algorithm, 56–57
    human respiratory motion, prediction
            of, 56
    prediction method, 57

signal history, selection of, 57
speed optimization, 57
Synchronized multileaf collimator
    for targets moving in 3D, 82
    for tongue and groove, 81
Synchrony Respiratory Tracking System, *2*
Synchrony system, 2
System integration testing, 134
System latency, 94
System validation testing, 134

T

Target alignment, *see* Couch-based target
    alignment
Target detection and tracking, theoretical
    aspects of, 13
    detection theory, implications of, 15
    image-based tracking, 15
    likelihood ratio test, 13–14, *14*
    matching cost functions, 16
        popular matching cost functions, 16–17
        robust matching cost functions, 17–18
    sensor modeling, 14–15, *15*, 15
    template selection and motion
        enhancement, 15–16, *16*, 17
    tracking and prediction, 18, *18*
        multiobject tracking, 19–20
        single object tracking, 18–19
Target motion detection, 21
Task retry, 133
Template selection and motion enhancement,
    15–16, *16*, *17*
Temporal adaptation (gating), 21
Temporal filtering and data association, 15
Testing process, 134
3D computed tomography, 66, 68, 69
3D conformal arc therapy, 42
3D marker location, 129
3D moving targets, DMLC control algorithms
    for, 81
    real-time synchronized MLC, 82–84
    synchronized MLC
        for targets moving in 3D, 82
        for tongue and groove, 81
3D surface imaging systems, 5
TMR, *see* Triple modular redundancy (TMR)
TomoDirect^SM delivery mode, 93, 94, 97,
    101, 103

Tomographic imaging, 33
TomoHelical^SM delivery mode, 93, 94, 97, 101,
    103
TomoTherapy® HiArt® system, *94*, 98, 117
TomoTherapy® treatment system,
    93–94, *94*
Tracking system, 130
Transient and low-frequency couch-based
    target alignment, 41, *42*
Transient faults, 131
Transverse motion
    compensation, of MAD strategy, 95, *95*
    components, 95
Treatment couch, in patient alignment, 39
Treatment planning
    and dosimetry, 108–111, *110*, *111*
    image processing in, 66, 68–69
    intratreatment motion adaption, 72–73, *73*
    for motion-compensated treatment,
        69–72, *69*, *70*, *71*, *72*
    for motion management via DMLC
        tracking, 77
    DMLC IMRT, 77–78
    DMLC tracking leaf-sequencing
        evolution, 78–79
    MLC tracking, of moving targets,
        88–90
    motion management, in VMAT, 86–87
    motion-optimized IMRT, 84–85
    1D moving and deforming targets,
        DMLC control algorithms for,
        79–81, *80*, *81*
    3D moving targets, DMLC control
        algorithms for, 81–84
    VMAT, interdependence of delivery
        parameters of, 90
    in radiation therapy, 65–66, *67*, *68*
Triple modular redundancy (TMR), 132, *133*
Tumor localization, *see* Real-time tumor
    localization
2D image-based tracking, 15
2D marker location, 129
Two-dimensional motion, tracking with, 35

U

Ultrasound, 1, 4
Unit testing, *see* Module testing
University Medical Center Utrecht, 4

V

Validation process, 134
Variable linear accelerator dose rate, DMLC
    control algorithms, 84–85
Varian Medical Systems Inc., 3, 5
Varian onboard imager (OBI) system, 1, 3, *3*
Verification process, 134
ViewRay, Inc., 115, 120
ViewRay™ system, 4, 115, 123–126, *125*
    gamma-ray IMRT, 119
        treatment plans, 120–121, *120*, *121*, *122*
    historical perspective, 115
        cobalt and conformal therapy, 115–116
        IGRT for radiotherapy, CT snapshots,
        118–119
        IMRT, 116–117
        MRI *vs.* CT imaging for intrafraction
        organ motion, 119
        patient and tumor positioning, 117–118
        technical developments, 116
    radiotherapy, and MRI, 121–126, *123*,
        124, *125*
Virginia Commonwealth University, 34
VisionRT Ltd., 5, *6*
Volumes of interest (VOI), 66, 69
Volumetric modulated arc therapy (VMAT), 77
    interdependence of delivery parameters
        of, 90
    motion management in, 86–87

W

Waiting Time, 130, *130*
Watchdog mechanism, 132, *132*
William Beaumont Hospital, 3

X

X-Ray 6D system, *see* Radiographic kV x-ray
    imaging system
X-ray imaging system, 1–2, 30, 31
Xsight® Lung Tracking System, 31

Z

Zero-mean version
    of matching cost functions, 17
    of robust matching cost function, 17

T - #0941 - 101024 - C8 - 276/216/8 - PB - 9781138374294 - Gloss Lamination